Lectures on
# STATISTICAL PHYSICS AND PROTEIN FOLDING

Lectures on
# STATISTICAL PHYSICS AND
# PROTEIN FOLDING

## Kerson Huang

*Massachusetts Institute of Technology*

**W🌐 World Scientific**

NEW JERSEY · LONDON · SINGAPORE · BEIJING · SHANGHAI · HONG KONG · TAIPEI · CHENNAI

*Published by*

World Scientific Publishing Co. Pte. Ltd.

5 Toh Tuck Link, Singapore 596224

*USA office:* 27 Warren Street, Suite 401-402, Hackensack, NJ 07601

*UK office:* 57 Shelton Street, Covent Garden, London WC2H 9HE

**British Library Cataloguing-in-Publication Data**
A catalogue record for this book is available from the British Library.

First published 2005
Reprinted 2006

ISBN-13  978-981-256-143-5
ISBN-10  981-256-143-9
ISBN-13  978-981-256-150-3 (pbk)
ISBN-10  981-256-150-1 (pbk)

Typeset by Stallion Press
Email: enquiries@stallionpress.com

Printed in Singapore

# Foreword

This book comprises a series of lectures given by the author at the Zhou Pei-Yuan Center for Applied Mathematics at Tsinghua University to introduce research in biology — specifically, protein structure — to individuals with a background in other sciences that also includes a knowledge in statistical physics. This is a timely publication, since the current perception is that biology and biophysics will undergo rapid development through applications of the principles of statistical physics, including statistical mechanics, kinetic theory, and stochastic processes.

The chapters begin with a good thorough introduction to statistical physics (Chapters 1–10). The presentation is somewhat tilted towards biological applications in the second part of the book (Chapters 11–16). Specific biophysical topics are then presented in this style while the general mathematical/physical principles, such as self-avoiding random walk and turbulence (Chapter 15), are further developed.

The discussion of "life process" begins with Chapter 11, where the basic topics of primary, secondary and tertiary structures are covered. This discussion ends with Chapter 16, in which working hypotheses are suggested for the basic principles that govern the formation and interaction of the secondary and tertiary structures. The author has chosen to avoid a more detailed discussion on empirical information; instead, references are given to standard publications. Readers who are interested in pursuing further in these directions are recommended to study *Mechanisms of Protein Folding* edited by Roger H. Pain (Oxford, 2000). Traditionally, the prediction of protein structure from its amino acid sequence has occupied the central position in the study of protein structure. Recently, however, there is a shift of emphasis towards the study of mechanisms. Readers interested in these general background information for better understanding of the present book are recommended to consult *Introduction to Protein Structure* by Carl Branden and John Tooze (Garland, 1999). Another strong point in this volume is the wide reproduction of key figures from these sources.

Protein structure is a complex problem. As is true with all complex issues, its study requires several different parallel approaches, which usually complement one another. Thus, we would expect that, in the long term, a better understanding of the mechanism of folding would contribute to the development of better methods of prediction. We look forward to the publication of a second edition of this volume in a few years in which all these new developments will be found in detail. Indeed, both of the two influential books cited above are in the second edition. We hope that this book will also play a similar influential role in the development of biophysics.

C.C. Lin

*Zhou Pei-Yuan Center for Applied Mathematics,*
*Tsinghua University, Beijing*

June 2004

# Contents

# Introduction

There is now a rich store of information on protein structure in various protein data banks. There is consensus that protein folding is driven mainly by the hydrophobic effect. What is lacking, however, is an understanding of specific physical principles governing the folding process. It is the purpose of these lectures to address this problem from the point of view of statistical physics. For background, the first part of these lectures provides a concise but relatively complete review of classical statistical mechanics and kinetic theory. The second part deals with the main topic.

It is an empirical fact that proteins of very different amino acid sequences share the same folded structure, a circumstance referred to as "convergent evolution." It other words, different initial states evolve towards the same dynamical equilibrium. Such a phenomenon is common in dissipative stochastic processes, as noted by C.C. Lin.[1] Some examples are the establishment of homogeneous turbulence, and the spiral structure of galaxies, which lead to the study of protein folding as a dissipative stochastic processes, an approach developed over the past year by the author in collaboration with Lin.

In our approach, we consider the energy balance that maintains the folded state in a dynamical equilibrium. For a system with few degrees of freedom, such as a Brownian particle, the balance between energy input and dissipation is relatively simple, namely, they are related through the fluctuation–dissipation theorem. In a system with many length scales, as a protein molecule, the situation is more complicated, and the input energy is dispersed among modes with different length scales, before being dissipated. Thus, energy

---

[1]C.C. Lin (2003). On the evolution of applied mathematics, *Acta Mech. Sin.* **19** (2), 97–102.

flows through the system along many different possible paths. The dynamical equilibrium is characterized by the most probable path.

- What is the source of the input energy?

The protein molecule folds in an aqueous solution, because of the hydrophobic effect. It is "squeezed" into shape by a fluctuating network of water molecules. If the water content is reduced, or if the temperature is raised, the molecule would become a random coil. The maintenance of the folded structure therefore requires constant interaction between the protein molecule and the water net. Water nets have vibrational frequencies of the order of 10 GHz. This lies in the same range as those of the low vibrational modes of the protein molecule. Therefore, there is resonant transfer of energy from the water network to the protein, in addition to the energy exchange due to random impacts. When the temperature is sufficiently low, the resonant transfer dominates over random energy exchange.

- How is the input energy dissipated?

The resonant energy transfer involves shape vibrations, and therefore occurs at the largest length scales of the protein molecule. It is then transferred to intermediate length scales through nonlinear couplings of the vibrational modes, most of which are associated with internal structures not exposed to the surface. There is thus little dissipation, until the energy is further dispersed down the ladder of length scales, until it reaches the surface modes associated with loops, at the smaller length scales of the molecule. Thus, there is energy cascade, reminiscent of that in the Kolmogorov theory of fully developed turbulence.

The energy cascade depends on the geometrical shape of the system, and the cascade time changes during the folding process. We conjecture that

*The most probable folding path is that which minimizes the cascade time.*

This principle may not uniquely determine the folded structure, but it would drive it towards a sort of "basin of attraction." This would provide a basis for convergent evolution, for the energy cascade blots out memory of the initial configuration after a few steps. A simple model in the Appendix illustrates this principle.

We shall begin with introductions to statistical methods, and basic facts concerning protein folding. The energy cascade will be discussed in the last two chapters.

For references on statistical physics, the reader may consult the following textbooks by the author:

K. Huang, *Introduction to Statistical Physics* (Taylor & Francis, London, 2001).

K. Huang, *Statistical Mechanics*, 2nd ed. (John Wiley & Sons, New York, 1987).

# Chapter 1

# Entropy

## 1.1. Statistical Ensembles

The purpose of statistical methods is to calculate the probabilities of occurrences of possible outcomes in a given process. We imagine that the process is repeated a large number of times $K$. If a specific outcome occurs $p$ number of times, then its probability of occurrence is defined as the limit of $p/K$, when $K$ tends to infinity. In such an experiment, the outcomes are typically distributed in the qualitative manner shown in Fig. 1.1, where the probability is peaked at some average value, with a spread characterized by the width of the distribution.

In statistical physics, our goal is to calculate the average values of physical properties of a system, such as correlation functions. The statistical approach is valid when fluctuations from average behavior are small. For most physical systems encountered in daily life,

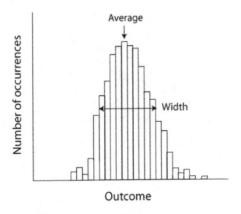

Fig. 1.1.    Relative probability distribution in an experiment.

1

fluctuations about average behavior are in fact small, due to the large number of atoms involved. This accounts for the usefulness of statistical methods in physics.

We calculate averages of physical quantities over a *statistical ensemble*, which consists of states of the system with assigned probabilities, chosen to best represent physical situations. By implementing such methods, we are able to derive the law of thermodynamics, and calculate thermodynamic properties, starting with an atomic description of matter. Historically, our theories fall into the following designations:

- *Statistical mechanics*, which deals with ensembles corresponding to equilibrium conditions;
- *Kinetic theory*, which deals with time-dependent ensembles that describe the approach to equilibrium.

Let us denote a possible state of a classical system by $s$. For definiteness, think of a classical gas of $N$ atoms, where the state of each atom is specified by the set of momentum and position vectors $\{\mathbf{p}, \mathbf{r}\}$. For the entire gas, $s$ stand for all the momenta and positions of all the $N$ atoms, and the phase space is $6N$-dimensional. The dynamical evolution is governed by the Hamiltonian $H(s)$, and may be represented by a trajectory in phase space, as illustrated symbolically in Fig. 1.2. The trajectory never intersects itself, since the solution to the equations of motion is unique, given initial conditions.

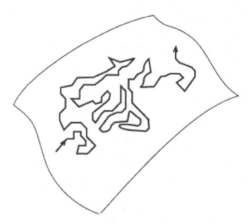

Fig. 1.2.   Symbolic representation of a trajectory in phase space.

It is exceedingly sensitive to initial conditions due to interactions. Two points near each other will initially diverge from each other exponentially in time, and the trajectory exhibits ergodic behavior: Given sufficient time, it will come arbitrarily close to any accessible point. After a short time, the trajectory becomes a spacing-filling tangle, and we can consider this as a distribution of points. This distribution corresponds to a statistical ensemble, which will continue to evolve towards an equilibrium ensemble.

There is a hierarchy of time scales, the shortest of which is set by the collision time, the average time interval between two successive atomic collisions, which is of the order of $10^{-10}$ s under standard conditions. Longer time scales are set by transport coefficients such as viscosity. Thus, a gas with arbitrary initial condition is expected to settle down to a state of local equilibrium in the order of $10^{-10}$ s, at which point a hydrodynamic description becomes valid. After a longer time, depending on initial conditions, the gas finally approaches a uniform equilibrium.

In the ensemble approach, we describe the distribution of points in phase space by a density function $\rho(s, t)$, which gives the relative probability of finding the state $s$ in the ensemble at time $t$. The ensemble average of a physical quantity $O(s)$ is then given by

$$\langle O \rangle = \frac{\sum_s O(s)\rho(s,t)}{\sum_s \rho(s,t)} \tag{1.1}$$

where the sum over states $s$ means integration over continuous variables. The equilibrium ensemble is characterized by a time-independent density function $\rho_{eq}(s) = \lim_{t\to\infty} \rho(s,t)$. Generally we assume that $\rho_{eq}(s)$ depends on $s$ only through the Hamiltonian: $\rho_{eq}(s) = \rho(H(s))$.

## 1.2. Microcanonical Ensemble and Entropy

The simplest equilibrium ensemble is a collection of equally weighted states, called the *microcanonical ensemble*. To be specific, consider an isolated macroscopic system with conserved energy. We assume that all states with the same energy $E$ occur with equal probability. Other parameters not explicitly mentioned, such as the number of particles and volume, are considered fixed properties. The phase-space volume occupied by the ensemble is

$$\Gamma(E) = \text{Number of states with energy } E \qquad (1.2)$$

This quantity is a measure of our uncertainty about the system, or the perceived degree of randomness. We define the entropy at a given energy as

$$S(E) = k_B \ln \Gamma(E) \qquad (1.3)$$

where $k_B$ is Boltzmann's constant, which specifies the unit of measurement. Since the phase-space volume of two independent systems is the product of the separate volumes, the entropy is additive.

The absolute temperature $T$ is defined by

$$\frac{1}{T} = \frac{\partial S(E)}{\partial E} \qquad (1.4)$$

For most systems, the number of states increases with energy, and therefore $T > 0$. For systems with energy spectrum bounded from above, however, the temperature can be negative, as illustrated in Fig. 1.3. In this case the temperature passes from $+\infty$ to $-\infty$ at the point of maximum entropy. A negative absolute temperature does not mean "colder than absolute zero," but "hotter than infinity," in the sense that any system in contact with it will draw energy from it. A negative temperature can in fact be realized experimentally in a spin system.

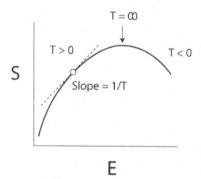

Fig. 1.3.   Temperature is related to the rate of increase of the number of states as energy increases.

## 1.3.  Thermodynamics

The energy difference between two equilibrium states is $dE = TdS$. Suppose the states are successive states of a system in a process in which no mechanical work was performed. Then the energy increase is due to heat absorption, by definition. Now we define the amount of heat absorbed in any process as

$$dQ = TdS \qquad (1.5)$$

even when mechanical work was done. If the amount of work done by the system is denoted be $dW$, we take the total change in energy as

$$dE = TdS - dW \qquad (1.6)$$

Heat is a form of disordered energy, since its absorption corresponds to an increase in entropy.

In classical thermodynamics, the quantities $dW$ and $dQ$ were taken as concepts derived from experiments. The *first law of thermodynamics* asserts that $dE = dQ - dW$ is an exact differential, while the *second law of thermodynamics* asserts that $dS = dQ/T$ is an exact differential. The point is that $dW$ and $dQ$ themselves are not exact differentials, but the combinations $dQ - dW$ and $dQ/T$ are exact.

In the statistical approach, $dE$ and $dS$ are exact differentials by construction. The content of the thermodynamic laws, in this view, is the introduction of the idea of heat.

## 1.4.  Principle of Maximum Entropy

An alternate form of the second law of thermodynamics states that the entropy of an isolated system never decreases. We can derive this principle using the definition of entropy in the microcanonical ensemble.

Consider a composite of two systems in contact with each other, labeled 1 and 2 respectively. For simplicity, let the systems be of the same type. The total energy $E = E_1 + E_2$ is fixed, but the energies of the component systems $E_1$ and $E_2$ can fluctuate. As illustrated in Fig. 1.4, $E_1$ can have a value below $E$, and $E_2$ is then determined

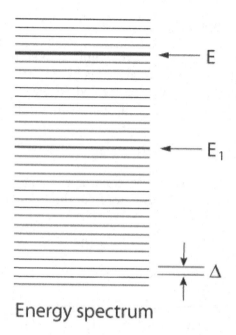

## Energy spectrum

Fig. 1.4.   The energy $E_1$ of a subsystem can range from the minimal energy to $E$. For a macroscopic system, however, it hovers near the value that maximizes its entropy.

as $E - E_1$. We have divided the energy spectrum into steps of level spacing $\Delta$, which denotes the resolution of energy measurements. The total number of accessible states is given by

$$\Gamma(E) = \sum_{E_0 < E_1 < E} \Gamma_1(E_1)\Gamma_2(E - E_1) \qquad (1.7)$$

where the sum extends over the possible values of $E_1$ in steps of $\Delta$. The total entropy is given by

$$S(E) = k_B \ln \sum_{E_0 < E_1 < E} \Gamma_1(E_1)\Gamma_2(E - E_1) \qquad (1.8)$$

For a macroscopic system, we will show that $E_1$ hovers near one value only — the value that maximizes its entropy.

Among the $E/\Delta$ terms in the sum, let the maximal term correspond to $E_1 = \bar{E}_1$. Since all terms are positive, the value of the sum lies between the largest term and $E/\Delta$ times the largest term:

$$
\begin{aligned}
k_B \ln\big[\Gamma_1(\bar{E}_1)\Gamma_2(E - \bar{E}_1)\big] \\
< S(E) < k_B \ln\big[\Gamma_1(\bar{E}_1)\Gamma_2(E - \bar{E}_1)\big] + \ln(E/\Delta)
\end{aligned}
\tag{1.9}
$$

In a macroscopic system of $N$ particles, we expect $S$ and $E$ both to be of order $N$. Therefore the last term on the right-hand side is of order $\ln N$, and may be neglected when $N \to \infty$. Thus

$$
S(E) = k_B \ln \Gamma_1(\bar{E}_1) + k_B \ln \Gamma_2(E - \bar{E}_1) + O(\ln N) \tag{1.10}
$$

Neglecting the last term, we have

$$
S(E) = S_1(\bar{E}_1) + S_2(\bar{E}_2) \tag{1.11}
$$

The principle of maximum entropy emerges when we compare (1.8) and (1.11). The former shows that the division of energy among subsystems have a range of possibilities. The latter indicates that, neglecting fluctuations, the energy is divided such as to maximize the entropy of the system.

As a corollary, we show that the condition for equilibrium between the subsystems is that their temperatures be equal. Maximizing $\ln[\Gamma_1(\bar{E}_1)\Gamma_2(E - \bar{E}_1)]$ with respect to $E_1$, we have

$$
\begin{aligned}
\frac{\partial}{\partial E_1}[\ln \Gamma_1(E_1) + \ln \Gamma_2(E - E_1)] = 0 \\
\frac{\partial}{\partial E_1} \ln \Gamma_1(E_1) - \frac{\partial}{\partial E_2} \ln \Gamma_2(E_2) = 0
\end{aligned}
\tag{1.12}
$$

This can be rewritten as

$$
\frac{\partial}{\partial E_1} S_1(E_1) = \frac{\partial}{\partial E_2} S_2(E_2) \tag{1.13}
$$

or

$$
T_1 = T_2 \tag{1.14}
$$

## 1.5. Example: Defects in Solid

Consider a lattice with $N$ sites, each occupied normally by one atom. There are $M$ possible interstitial locations where atoms can be misplaced, and it costs an energy $\Delta$ to misplace an atom, as illustrated in Fig. 1.5. Assume $N$, $M \to \infty$, and the number of displaced atoms $n$ is a small fraction of $N$. Calculate the thermodynamic properties of this system. The given macroscopic parameters are $N, M, n$. The energy is

$$E = n\Delta \qquad (1.15)$$

The number of states in a microcanonical ensemble is

$$\Gamma(E) = \left[\frac{N!}{n!(N-n)!}\right]\left[\frac{M!}{n!(M-n)!}\right] \qquad (1.16)$$

The first factor is the number of ways to choose the $n$ atoms to be removed from $N$ sites, and the second factor is the number of ways to place the $n$ atoms on the $M$ interstitials. We can use Stirling's approximation for the factorials:

$$\ln N! \approx N \ln N - N \qquad (1.17)$$

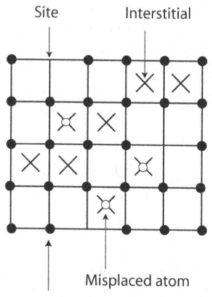

Fig. 1.5.   Model of defects in a solid.

The entropy of the system is then

$$\frac{S(E)}{k_B} = \ln \Gamma(E) = n \ln \frac{N}{n} - (N - n) \ln \left(1 - \frac{n}{N}\right)$$

$$+ n \ln \frac{M}{n} - (M - n) \ln \left(1 - \frac{n}{M}\right) \tag{1.18}$$

The temperature is given through

$$\frac{1}{k_B T} = \frac{1}{k_B} \frac{\partial S(E)}{\partial E} = \frac{\partial \ln \Gamma(E)}{\partial E} = \frac{1}{\Delta} \frac{\partial \ln \Gamma(E)}{\partial n} \tag{1.19}$$

This leads to

$$\frac{\Delta}{k_B T} = \frac{\partial}{\partial n} \ln \Gamma(E) = \ln \left(\frac{N}{n} - 1\right) + \ln \left(\frac{M}{n} - 1\right) \tag{1.20}$$

Exponentiating both sides, we have

$$\frac{n^2}{(N - n)(M - n)} = \exp\left(-\frac{\Delta}{k_B T}\right) \tag{1.21}$$

The low- and high-temperature limits are

$$\begin{aligned} n &\approx \sqrt{NM} \exp(-\Delta/2k_B T) & (k_B T \ll \Delta) \\ \frac{1}{n} &\approx \frac{1}{N} + \frac{1}{M} & (k_B T \gg \Delta) \end{aligned} \tag{1.22}$$

As a model for defects in a solid, we set $N = M$, and $\Delta = 1\,\text{eV}$. Then

$$\frac{n}{N} \approx \exp(-\Delta/2k_B T) \tag{1.23}$$

For $T = 300\,\text{K}$: $n/N \approx 2 \times 10^{-9}$.
For $T = 1000\,\text{K}$: $n/N \approx 2.5 \times 10^{-3}$.

# Chapter 2

# Maxwell–Boltzmann Distribution

## 2.1. Classical Gas of Atoms

For the macroscopic behavior of a classical gas of atoms, we are not interested in the precise coordinates $\{\mathbf{p}, \mathbf{r}\}$ of each atom. All we need to know is the number of atoms with a given $\{\mathbf{p}, \mathbf{r}\}$, to a certain accuracy. Accordingly, we group the values of $\{\mathbf{p}, \mathbf{r}\}$ into cells of size $\Delta\tau$ corresponding to a given energy tolerance. The cells are assumed to be sufficiently large to contain a large number of atoms, and yet small enough to be considered infinitesimal on a macroscopic scale.

Label the cells by $\lambda = 1, \ldots, K$. The positions and momenta in cell $\lambda$ have unresolved values $\{\mathbf{r}_\lambda, \mathbf{p}_\lambda\}$, and the corresponding kinetic energy is $\epsilon_\lambda = \mathbf{p}_\lambda^2/2m$. For a very dilute gas, we neglect the interatomic interactions, and take the total energy $E$ to be the sum of kinetic energies over all the cells.

The number of atoms in cell $\lambda$ is called the *occupation number* $n_\lambda$. A set of occupation numbers $\{n_1, n_2, \ldots\}$ is called a *distribution*. Since there are $N$ atoms with total energy $E$, we have the conditions

$$\sum_\lambda n_\lambda = N$$

$$\sum_\lambda n_\lambda \epsilon_\lambda = E \tag{2.1}$$

The number of states corresponding to the distribution $\{n_1, n_2, \ldots\}$ is the number of permutations of $N$ particles that interchange particles in different cells:

$$\frac{N!}{n_1! n_2! \cdots n_K!} \tag{2.2}$$

The phase-space volume of the microcanonical ensemble is obtained, up to a multiplicative factor, by summing the above over all allowable distributions, except that the factor $N!$ is to be omitted:

$$\Gamma(E, V) = \sum_{\{n_\lambda\}} \Omega(n_1, n_2, \ldots)$$

$$\Omega(n_1, n_2, \ldots) = \frac{1}{n_1! n_2! \cdots n_K!} \tag{2.3}$$

where the sum $\sum_{\{n_\lambda\}}$ extends over all possible sets $\{n_\lambda\}$ that satisfy the constraints (2.1).

The factor $N!$ was omitted according to a recipe called the "correct Boltzmann counting", which is dictated by correspondence with quantum mechanics. It has no effect on processes in which $N$ is kept constant, but is essential to avoid inconsistencies when $N$ is variable. The recipe only requires that we omit a factor proportional to $N!$. Consequently, the phase-space volume is determined only up to an arbitrary constant factor.

## 2.2.  The Most Probable Distribution

The entropy of the system is, up to an arbitrary additive constant,[1]

$$S(E, V) = k \ln \sum_{\{n_\lambda\}} \Omega(n_1, n_2, \ldots) \tag{2.4}$$

This is expected to be of order $N$. By an argument used in the last chapter, we only need to keep the largest term in the sum above:

$$S(E, V) = k \ln \Omega(\bar{n}_1, \bar{n}_2, \ldots) + O(\ln N) \tag{2.5}$$

where the distribution $\{\bar{n}_\lambda\}$ maximizes $\Omega$, and is called the *most probable distribution*. That is, $\delta \ln \Omega = 0$ under the variation $n_\lambda \to \bar{n}_\lambda + \delta n_\lambda$, with the constraints

$$\sum_\lambda \delta n_\lambda = 0$$

$$\sum_\lambda \epsilon_\lambda \delta n_\lambda = 0 \tag{2.6}$$

---

[1]From now on, we denote Boltzmann's constant by $k$ instead of $k_B$, when no confusion arises.

These are taken into account by introducing Lagrange multipliers. That is, we consider

$$\delta\left(\ln\Omega + \alpha\sum_\lambda n_\lambda - \beta\sum_\lambda \epsilon_\lambda n_\lambda\right) = 0 \qquad (2.7)$$

where each $n_\lambda$ is to be varied independently, and $\alpha$ and $\beta$ are fixed parameters called Lagrange multipliers. We determine $\alpha$ and $\beta$ afterwards to satisfy (2.1).

Using the Stirling approximation, we have

$$\ln\Omega = -\sum_\lambda \ln n_\lambda! \approx -\sum_\lambda n_\lambda \ln n_\lambda + \sum_\lambda n_\lambda \qquad (2.8)$$

hence

$$\sum_\lambda (\ln n_\lambda - \alpha + \beta\epsilon_\lambda)\delta n_\lambda = 0 \qquad (2.9)$$

Since the $\delta n_\lambda$ are arbitrary and independent, we must have $\ln n_\lambda = \alpha - \beta\epsilon_\lambda$. Thus the most probable distribution is

$$\bar{n}_\lambda = \alpha e^{-\beta\epsilon_\lambda} \qquad (2.10)$$

This is called the Maxwell–Boltzmann distribution.

## 2.3. The Distribution Function

We now "zoom out" to a macroscopic view, in which the cell size becomes very small. The cell label $\lambda$ becomes $\{\mathbf{p}, \mathbf{r}\}$, and $\Delta\tau$ becomes an infinitesimal volume element:

$$\lambda \to \{\mathbf{p}, \mathbf{r}\}$$
$$\Delta\tau \to \frac{d^3p\, d^3r}{h^3} \qquad (2.11)$$
$$\sum_\lambda \to \int \frac{d^3p\, d^3r}{h^3}$$

where $h$ is a constant specifying the units, chosen to be Planck's constant for correspondence with quantum mechanics. The occupation number becomes infinitesimal:

$$n_\lambda \rightarrow f(\mathbf{p}, \mathbf{r}) \frac{d^3 p \, d^3 r}{h^3} \tag{2.12}$$

where $f(\mathbf{p}, \mathbf{r})$ is called the *distribution function*. The most probable distribution corresponds to

$$f(\mathbf{p}) = Ce^{-\beta \mathbf{p}^2 / 2m} \tag{2.13}$$

which is independent of $\mathbf{r}$ in the absence of external potential. The constraints (2.1) become

$$\int \frac{d^3 p}{h^3} f(\mathbf{p}) = \frac{N}{V} = n$$
$$\frac{1}{2m} \int \frac{d^3 p}{h^3} \mathbf{p}^2 f(\mathbf{p}) = \frac{E}{V} \tag{2.14}$$

where $V$ is the volume of the system, and $n$ is the particle density. We need the integrals

$$\int_0^\infty dx \, x^2 e^{-bx^2} = \frac{\sqrt{\pi}}{4} b^{-3/2}$$
$$\int_0^\infty dx \, x^4 e^{-bx^2} = \frac{3\sqrt{\pi}}{8} b^{-5/2} \tag{2.15}$$

The constraints are then satisfied with

$$\alpha = n\lambda^3$$
$$\beta = \frac{3N}{2E} \tag{2.16}$$

where

$$\lambda = \sqrt{\frac{2\pi\hbar^2 \beta}{m}} \tag{2.17}$$

is a parameter of dimension length, with $\hbar = h/2\pi$. It follows that $\beta = (kT)^{-1}$, and $\lambda$ is the *thermal wavelength*, the deBroglie

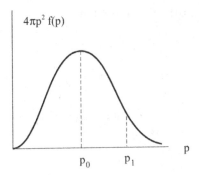

Fig. 2.1. Maxwell–Boltzmann distribution of magnitude of momentum.

wavelength of a particle of energy $kT$. This completes the determination of the Maxwell–Boltzmann distribution function.

The physical interpretation of the distribution function is

$$f(\mathbf{p}) \, d^3p$$
$$= \text{Probability of finding an atom with momentum } \mathbf{p} \text{ within } d^3p$$
$$(2.18)$$

The probability density $4\pi p^2 f(\mathbf{p})$ is qualitatively sketched in Fig. 2.1. This gives the probability per unit volume of finding $|\mathbf{p}|$ between $p$ and $p + dp$. The area under the curve is the density of the gas $n$. The maximum of the curve corresponds to the "most probable momentum" $p_0 = mv_0$, which gives the "most probable velocity"

$$v_0 = \sqrt{\frac{2}{m\beta}} \qquad (2.19)$$

The root-mean-square average of the momentum $p_1 = mv_1$ gives the root-mean-square velocity

$$v_1 = \sqrt{\frac{3}{m\beta}} \qquad (2.20)$$

## 2.4.   Thermodynamic Properties

The entropy can be obtained from (2.5):

$$S = k \ln \Omega\{\bar{n}_i\} = -k \sum_{\lambda} \bar{n}_\lambda \ln \bar{n}_\lambda$$

$$= -\frac{kV}{h^3} \int d^3 p\, f(\mathbf{p}) \ln f(\mathbf{p}) \qquad (2.21)$$

This leads to

$$\frac{S}{Nk} = -\ln(\lambda^3 n) = \ln \frac{V}{N} + \frac{3}{2} \ln\left(\frac{E}{N}\right) \qquad (2.22)$$

up to an additive constant. Inverting this relation gives the internal energy as a function of $S$ and $V$:

$$\frac{E(S,V)}{N} = c_0 \left(\frac{N}{V}\right)^{2/3} \exp\left(\frac{2}{3}\frac{S}{Nk}\right) \qquad (2.23)$$

where $c_0$ is a constant.

The absolute temperature is given by

$$T = \frac{\partial E(S,V)}{\partial S} = \frac{2}{3k}\frac{E}{N} \qquad (2.24)$$

Thus

$$\frac{E}{N} = \frac{3}{2}kT \qquad (2.25)$$

Comparison with (2.16) shows $\beta = (kT)^{-1}$. This formula expresses the *equipartition of energy,* namely, the thermal energy residing in each translational degree of freedom is $\frac{1}{2}kT$.

A formula for the pressure can be obtained from the first law $dE = TdS - PdV$, by setting $dS = 0$:

$$P = -\frac{\partial E(S,V)}{\partial V} = \frac{2}{3}\frac{E}{V} = \frac{NkT}{V} \qquad (2.26)$$

which is the ideal gas law.

# Chapter 3

# Free Energy

## 3.1. Canonical Ensemble

We have used the microcanonical ensemble to describe an isolated system. However, most systems encountered in the laboratory are not isolated. What would be the ensemble appropriate for such cases? The answer is found within the microcanonical ensemble, by examining a small part of an isolated system. We focus our attention on the small subsystem, and regard the rest of the system as a "heat reservoir", with which the subsystem exchanges energy.

Label the small system 1, and the heat reservoir 2, as illustrated schematically in Fig. 3.1. Working in the microcanonical ensemble for the whole system, we will find that system 1 is described by an ensemble of fixed temperature instead of fixed energy, and this is called the canonical ensemble.

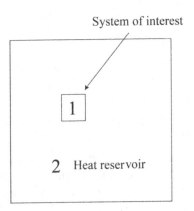

Fig. 3.1.   We focus our attention on the small subsystem 1. The rest of the system acts as a heat reservoir with a fixed temperature.

The total number of particles and total energy are sums of those in the two systems:

$$N = N_1 + N_2$$
$$E = E_1 + E_2 \tag{3.1}$$

where

$$N_2 \gg N_1$$
$$E_2 \gg E_1 \tag{3.2}$$

Assuming that both systems are macroscopically large, we have neglected interaction energies across the boundaries of the system. We keep $N_1$ and $N_2$ separately fixed, but allow $E_1$ and $E_2$ to fluctuate. In other words, the boundaries between the two subsystems allow energy exchange, but not particle exchange.

We wish to find the phase-space density $\rho_1(s_1)$ for system 1 in its own phase space. This is proportional to the probability of finding system 1 in state $s_1$, regardless of the state of system 2. It is thus proportional to the phase-space volume of system 2 in its own phase space, at energy $E_2$. The proportionality constant being unimportant, we take

$$\rho_1(s_1) = \Gamma_2(E_2) = \Gamma_2(E - E_1) \tag{3.3}$$

Since $E_1 \ll E$, we shall expand $\Gamma_2(E - E_1)$ in powers of $E_1$ to lowest order. It is convenient to expand $k \ln \Gamma_2$, which is the entropy of system 2:

$$k \ln \Gamma_2(E - E_1) = S_2(E - E_1)$$

$$= S_2(E) - E_1 \left. \frac{\partial S_2(E')}{\partial E'} \right|_{E'=E} + \cdots$$

$$\approx S_2(E) - \frac{E_1}{T} \tag{3.4}$$

where $T$ is the temperature of system 2. This relation becomes exact in the limit when system 2 becomes infinitely larger than system 1. It then becomes a heat reservoir with given temperature $T$. The density function for system 1 is therefore

$$\rho_1(s_1) = e^{S_2(E)/k} e^{-E_1/kT} \tag{3.5}$$

The first factor is a constant, which can be dropped by redefining the normalization. In the second factor, the energy of the system can be replaced by the Hamiltonian:

$$E_1 = H_1(s_1) \tag{3.6}$$

Since we shall no longer refer to system 2, subscripts are no longer necessary, and will be omitted. Thus, the density function for a system held at temperature $T$ is

$$\rho(s) = e^{-\beta H(s)} \tag{3.7}$$

where $H(s)$ is the Hamiltonian of the system, and $\beta = 1/kT$. This defines the *canonical ensemble*.

It is useful to introduce the *partition function*:

$$Q(T, V) \equiv \sum_s e^{-\beta H(s)} \tag{3.8}$$

where the sum extends over all states $s$ of the system, each weighted by the Boltzmann factor

$$e^{-\text{Energy}/kT} \tag{3.9}$$

Compared to the microcanonical ensemble, the constraint of fixed energy has been relaxed, as illustrated schematically in Fig. 3.2. However, the thermodynamic properties resulting from these two ensembles are equivalent. This is because the energy in the canonical ensemble fluctuates about a mean value, and the fluctuations are negligible for a macroscopic system, as we now show.

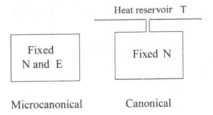

Fig. 3.2. Schematic representations of microcanonical ensemble and canonical ensemble.

## 3.2. Energy Fluctuations

The mean energy $U$ in the canonical ensemble is given by the ensemble average of the Hamiltonian:

$$U = \frac{\sum_s H e^{-\beta H}}{\sum_s e^{-\beta H}} = -\frac{\partial}{\partial \beta} \ln \sum_s e^{-\beta H} \qquad (3.10)$$

Differentiating with respect to $\beta$, we have

$$\frac{\partial U}{\partial \beta} = -\frac{\sum_s H^2 e^{-\beta H}}{\sum_s e^{-\beta H}} + \frac{\left(\sum_s H e^{-\beta H}\right)^2}{\left(\sum_s e^{-\beta H}\right)^2} = -\langle H^2 \rangle + \langle H \rangle^2 \qquad (3.11)$$

We can rewrite

$$\frac{\partial U}{\partial \beta} = \frac{\partial U}{\partial T} \frac{\partial T}{\partial \beta} = -kT^2 \frac{\partial U}{\partial T} = -kT^2 C_V \qquad (3.12)$$

where $C_V$ is the heat capacity at constant volume. Thus

$$\langle H^2 \rangle - \langle H \rangle^2 = kT^2 C_V \qquad (3.13)$$

For macroscopic systems the left side is of order $N^2$, while the right side is of order $N$. Energy fluctuations therefore become negligible when $N \to \infty$.

## 3.3. The Free Energy

We can examine energy fluctuation in more detail, by rewriting the partition function (3.8) as an integral over energy. To do this, we insert into the sum a factor of identity in the form

$$\int dE \delta(E - H(s)) = 1 \qquad (3.14)$$

Thus,

$$Q(T,V) = \sum_s \int dE \delta(E - H(s)) e^{-\beta H(s)} \qquad (3.15)$$

Interchanging the order of integration and summation, we can write

$$Q = \int dE e^{-\beta E} \Gamma(E)$$

$$\Gamma(E) = \sum_s \delta(H(s) - E)$$

$$(3.16)$$

The integrand is the product of the Boltzmann factor $e^{-\beta E}$, which is a decreasing function, with the number of states $\Gamma(E)$, which is increasing. Thus it is peaked at some value of the energy. For macroscopic systems, the factors involved change rapidly with energy, making the peak extremely sharp.

We note that $\Gamma(E)$ is the phase-space volume of a microcanonical ensemble of energy $E$, and thus related to the entropy of the system by $S(E) = k \ln \Gamma(E)$. Thus

$$Q = \int dE e^{-\beta[E-TS(E)]} = \int dE e^{-\beta A(E)} \tag{3.17}$$

where

$$A(E) \equiv E - TS(E) \tag{3.18}$$

is the free energy at energy $E$. The term $TS$ represents the part of the energy residing in random thermal motion. Thus, the free energy represents the part of the energy available for performing work.

The integrand in (3.17) is peaked at $E = \bar{E}$ where $A(E)$ is at a minimum:

$$\left.\frac{\partial A}{\partial E}\right|_{E=\bar{E}} = \left[1 - T\frac{\partial S}{\partial E}\right]_{E=\bar{E}} = 0 \tag{3.19}$$

or

$$\left(\frac{\partial S}{\partial E}\right)_{E=\bar{E}} = \frac{1}{T} \tag{3.20}$$

In other words, $\bar{E}$ is the energy at which we have the thermodynamic relation between entropy and temperature. The second derivative of $A(E)$ gives[1]

$$\frac{\partial^2 A}{\partial E^2} = -T\frac{\partial^2 S}{\partial E^2} = \frac{1}{TC_V} \tag{3.21}$$

Now expand $A(E)$ about the minimum:

$$A(E) = A(\bar{E}) + \frac{1}{2TC_V}(E - \bar{E})^2 + \cdots \tag{3.22}$$

Neglecting the higher-order terms, we have

$$Q = e^{-\beta A(\bar{E})} \int dE\, e^{-(E-\bar{E})^2/(2kT^2C_V)} \tag{3.23}$$

Since $C_V$ is of order $N$, the integrand is very sharply peaked at $E = \bar{E}$, as illustrated in Fig. 3.3. The width of the peak is $\sqrt{kT^2C_V}$, which is the root-mean-square fluctuation obtained earlier by a different method. Since the peak is very sharp, we can perform the integration over energy by extending the limits of integration from $-\infty$ to $\infty$, and obtain

$$\ln Q = -\beta A(\bar{E}) + \frac{1}{2}\ln\left(2\pi kT^2C_V\right) \tag{3.24}$$

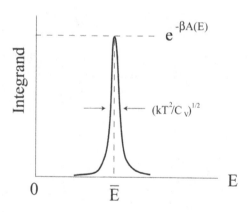

Fig. 3.3.   When the partition function is expressed as an integral over energy, the integrand is sharply peaked at a value corresponding to a minimum of the free energy.

---

[1]The last equality comes from $\partial S/\partial E = -1/T$, hence $\partial^2 S/\partial E^2 = T^{-2}\partial T/\partial E = 1/(T^2 C_V)$.

In the thermodynamic limit, the first term is of order $N$, while the second term is of order $\ln N$, and can be neglected.

In summary, we have derived two thermodynamic results:

- In the canonical ensemble with given temperature $T$ and volume $V$, thermodynamic functions can be obtained from the free energy $A(V, T)$, via the connection

$$\sum_s e^{-\beta H(s)} = e^{-\beta A(V,T)} \qquad (3.25)$$

- At fixed temperature and volume, thermodynamic equilibrium corresponds to the state of minimum free energy.

## 3.4. Maxwell's Relations

All the thermodynamic functions of a system can be derived from a single function. We have seen that those of an isolated system can be derived from the energy $U(S, V)$. This must be expressed as a function of $S$ and $V$, for then we obtain all other properties though use of the first law with $S$ and $V$ appearing as independent variables:

$$dU = TdS - PdV$$
$$T = \frac{\partial U}{\partial S} \quad P = -\frac{\partial U}{\partial V} \qquad (3.26)$$

where the last two formulas are called *Maxwell relations*. For other types of processes, we use different functions:

- Constant $T, V$: Use the free energy $A(T, V) = U - TS$:

$$dA = -SdT - PdV$$
$$S = -\frac{\partial A}{\partial T} \quad P = -\frac{\partial A}{\partial V} \qquad (3.27)$$

- Constant $T, P$; Use the Gibbs potential $G(P, T) = A + PV$:

$$dG = -SdT + VdP$$
$$S = -\frac{\partial G}{\partial T} \quad V = \frac{\partial G}{\partial P} \qquad (3.28)$$

Fig. 3.4.  Each quantity at the center of a row or column is flanked by its natural variables. The partial derivative with respect to one of the variables, with the other held fixed, is arrived at by following the diagonal line originating from that variable. Attach a minus sign if you go against the arrow.

- Constant $P, S$: Use the enthalpy $H(P, S) = U + PV$:

$$dH = TdS + VdP$$
$$T = \frac{\partial H}{\partial S} \quad V = \frac{\partial H}{\partial P} \tag{3.29}$$

All the Maxwell relations can be conveniently summarized as in Fig. 3.4.

## 3.5.  Example: Unwinding of DNA

The unwinding of a double-stranded DNA molecule is like unraveling a zipper. The DNA has $N$ links, each of which can be in one of two states: a closed state with energy 0, and an open state with energy $\Delta$. A link can be opened only if all the links to its left are already open, as illustrated in Fig. 3.5. Due to thermal fluctuations, links will spontaneously open and close. What is the average number of open links?

The possible states are labeled by the number of open links $n = 0, 1, 2, \ldots, N$. The energy with $n$ open links is $E_n = n\Delta$. The

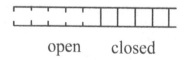

open       closed

Fig. 3.5.  Zipper model of DNA.

partition function is

$$Q_N = \sum_{n=0}^{N} e^{-\beta n \Delta} = \frac{1 - e^{-\beta(\bar{N}+1)\Delta}}{1 - e^{-\beta\Delta}}$$

The average number of open links is

$$\bar{n} = \frac{\sum_{n=0}^{N} n e^{-\beta n \Delta}}{\sum_{n=0}^{N} e^{-\beta n \Delta}} = -\frac{1}{\Delta} \frac{\partial \ln Q_N}{\partial \beta}$$

$$= \frac{e^{-\beta\Delta}}{1 - e^{-\beta\Delta}} - \frac{(N+1)e^{-\beta(\bar{N}+1)\Delta}}{1 - e^{-\beta(\bar{N}+1)\Delta}}$$

At low temperatures $\beta\Delta \gg 1$, and there are few open links:

$$\bar{n} \approx e^{-\beta\Delta}$$

At high temperatures $\beta\Delta \ll 1$, almost all links are open:

$$\bar{n} \approx n$$

# Chapter 4

# Chemical Potential

## 4.1. Changing the Particle Number

The chemical potential $\mu$ is the energy required for adding one particle to the system. When the total number of particles $N$ varies, the first law of thermodynamics is generalized to

$$dU = TdS - PdV + \mu dN \tag{4.1}$$

or, equivalently,

$$dA = -SdT - PdV + \mu dN \tag{4.2}$$

Thus

$$\mu = \frac{\partial A(V, T, N)}{\partial N} \tag{4.3}$$

A related variable is the fugacity

$$z = e^{\mu/kT} \tag{4.4}$$

Some useful thermodynamic relations follow from the property

$$A(V, T, N) = Na(v, T) \tag{4.5}$$

where $a(v, T)$ is the free energy per particle at specific volume $v = V/N$. Thus

$$\mu = a(v, T) + N\frac{\partial}{\partial N}a(v, T) \tag{4.6}$$

Since

$$N\frac{\partial a}{\partial N} = N\frac{\partial a}{\partial v}\frac{\partial v}{\partial N} = -v\frac{\partial a}{\partial v} = Pv \tag{4.7}$$

where $P$ is the pressure, we have

$$\mu = a + Pv \qquad (4.8)$$

Differentiating this with respect to $v$, we obtain $\partial\mu/\partial v = v\partial P/\partial v$, which implies $(\partial P/\partial v)\,(\partial v/\partial\mu) = 1/v$, or

$$\frac{\partial P}{\partial\mu} = \frac{1}{v} \qquad (4.9)$$

## 4.2.  Grand Canonical Ensemble

We can remove the restriction to a fixed number of particles in the canonical ensemble, by allowing the system to exchange particles with a particle reservoir of given chemical potential. The number of particles will then fluctuate about a mean value. The resulting ensemble is called the *grand canonical ensemble*. It has the advantage of being applicable to situations in which the particle number is variable, which is almost always the case for a macroscopic system. This is illustrated in Fig. 4.1. The grand partition function is defined as a sum over the partition functions $Q_N$ for different particle numbers $N$, weighted by the $N$th power of the fugacity:

$$\mathcal{Q}(z, V, T) = \sum_{N=0}^{\infty} z^N Q_N(V, T) \qquad (4.10)$$

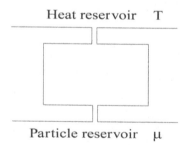

Fig. 4.1.   In the grand canonical ensemble, the system can exchange particles with a reservoir of given chemical potential. As in the canonical ensemble, it exchanges energy with a reservoir of given temperature.

In the macroscopic limit we expect

$$\frac{1}{V} \ln \mathcal{Q}(z, V, T) \xrightarrow[V \to \infty]{} \text{Finite limit} \tag{4.11}$$

The average number of particles is given by the ensemble average

$$\bar{N} = \frac{\sum N z^N Q_N}{\sum z^N Q_N} = z \frac{\partial}{\partial z} \ln \mathcal{Q}(z, V, T) \tag{4.12}$$

The mean-square fluctuation can be obtained by differentiating again with respect to $z$:

$$z \frac{\partial}{\partial z} z \frac{\partial}{\partial z} \ln \mathcal{Q}(z, V, T) = \frac{\sum N^2 z^N Q_N}{\sum z^N Q_N} - \left[ \frac{\sum N z^N Q_N}{\sum z^N Q_N} \right]^2$$
$$= \overline{N^2} - \bar{N}^2$$

In terms of the chemical potential we can write

$$z \frac{\partial}{\partial z} = z \frac{\partial \mu}{\partial z} \frac{\partial}{\partial \mu} = kT \frac{\partial}{\partial \mu} \tag{4.13}$$

Thus

$$\overline{N^2} - \bar{N}^2 = (kT)^2 \frac{\partial^2}{\partial \mu^2} \ln \mathcal{Q}(z, V, T) \tag{4.14}$$

Dividing both sides by $V^2$, we have the density fluctuation

$$\overline{n^2} - \bar{n}^2 = \frac{(kT)^2}{V^2} \frac{\partial^2}{\partial \mu^2} \ln \mathcal{Q}(z, V, T) \tag{4.15}$$

Assuming (4.11), we see it vanishes like $V^{-1}$ in the thermodynamic limit. This makes the grand canonical ensemble equivalent to the canonical ensemble.

Like the energy in the canonical ensemble, the particle number here fixes itself in a macroscopic system, except in the neighborhood of a phase transition, which we shall discuss later.

## 4.3.  Thermodynamics

Assuming that the number fluctuation is vanishingly small, we need to keep only the largest term in the sum over $N$:

$$\ln \mathcal{Q}(z, V, T) \approx \ln\left[e^{\beta N\mu/kT}Q_{\bar{N}}(V,T)\right] = \beta N\mu + \ln Q_N(V,T) \quad (4.16)$$

where $\beta = 1/kT$, and $N$ denotes the average number of particles. Taking $\ln Q_N = -\beta A_N = -\beta N a$, we have

$$\ln \mathcal{Q}(z, V, T) = \frac{N}{kT}[\mu - a(v,T)] = \frac{PV}{kT} \quad (4.17)$$

where we have used (4.8) in the last step. Combining with (4.12) leads to the parametric equations

$$\frac{P}{kT} = \frac{1}{V}\ln \mathcal{Q}(z, V, T)$$

$$n = \frac{1}{V}z\frac{\partial}{\partial z}\ln \mathcal{Q}(z, V, T) \quad (4.18)$$

from which $z$ should be eliminated to obtain the pressure as a function of density and temperature.

## 4.4.  Critical Fluctuations

From (4.15) and (4.17), we have

$$\overline{n^2} - \bar{n}^2 = \frac{kT}{V}\frac{\partial^2 P}{\partial \mu^2} \quad (4.19)$$

Using (4.9), we have

$$\frac{\partial^2 P}{\partial \mu^2} = \frac{\partial}{\partial \mu}\frac{1}{v} = -\frac{1}{v^2}\frac{1}{\partial \mu/\partial v} = -\frac{1}{v^3 \partial P/\partial v} \quad (4.20)$$

Introducing the isothermal compressibility

$$\kappa_T = -\frac{1}{v}\left(\frac{\partial v}{\partial P}\right)_T \quad (4.21)$$

we obtain the fractional density fluctuation:

$$\frac{\overline{n^2} - \bar{n}^2}{\bar{n}^2} = \frac{kT\kappa_T}{V} \tag{4.22}$$

This vanishes when $V \to \infty$, except when $\kappa_T \to \infty$, and the latter happens at the critical point of a phase transition. In reality, the fluctuation does not diverge, for the atomic structure of matter acts as a cutoff. But it becomes extremely large.

The mean-square fluctuation of density in an atomic system is proportional to the scattering cross-section of light. This relation has observable consequences, as in the blue of the sky. The large fluctuation at the critical point leads to *critical opalescence*. In $CO_2$, the intensity of scattered light increases a million fold at its critical point at $T = 304\,\mathrm{K}$, $P = 74$ atm, and the normally transparent liquid turns milky white.

## 4.5. Example: Ideal Gas

The partition function is

$$Q_N(T, V) = \int \frac{d^{3N}p\, d^{3N}r}{N! h^{3N}} \exp\left(-\frac{1}{2mkT}\sum_{i=1}^{N} \mathbf{p}_i^2\right) \tag{4.23}$$

$$= \frac{V^N}{N!}\left(\int_{-\infty}^{\infty} \frac{dp}{h}\, e^{-\beta p^2/2m}\right)^{3N} = \frac{1}{N!}\left(\frac{V}{\lambda^3}\right)^N \tag{4.24}$$

where the phase-space volume element is divided by $N!$ in accordance with correct Boltzmann counting, and

$$\lambda = \sqrt{2\pi\hbar^2/mkT} \tag{4.25}$$

In the canonical ensemble we obtain all thermodynamic information from the free energy

$$A(V, T) = -kT \ln Q_N(T, V) = NkT\left[\ln\left(n\lambda^3\right) - 1\right] \tag{4.26}$$

where $n = N/V$, and we have used the Stirling approximation $\ln N! \approx N \ln N - N$. We obtain

$$\text{Chemical potential:} \quad \mu = \left(\frac{\partial A}{\partial N}\right)_{V,T} = kT \ln\left(n\lambda^3\right) \qquad (4.27)$$

$$\text{Entropy:} \quad S = -\left(\frac{\partial A}{\partial T}\right)_V = Nk\left[\frac{5}{2} - \ln\left(n\lambda^3\right)\right] \qquad (4.28)$$

$$\text{Pressure:} \quad P - \left(\frac{\partial A}{\partial V}\right)_T = \frac{NkT}{V} \qquad (4.29)$$

In the grand canonical ensemble, the grand partition function is given by

$$\mathcal{Q}(z, V, T) = \sum_{N=0}^{\infty} \frac{z^N}{N!}\left(\frac{V}{\lambda^3}\right)^N = \exp\frac{zV}{\lambda^3} \qquad (4.30)$$

Thus

$$\frac{1}{V}\ln\mathcal{Q}(z, V, T) = \frac{z}{\lambda^3} \qquad (4.31)$$

The parametric equation of state is

$$\begin{aligned} \frac{P}{kT} &= \frac{z}{\lambda^3} \\ n &= \frac{z}{\lambda^3} \end{aligned} \qquad (4.32)$$

from which we recover the pressure and chemical potential obtained earlier.

# Chapter 5

# Phase Transitions

A phase transition is an abrupt change in thermodynamic behavior, associated with a discontinuity in some thermodynamic function. At constant pressure, it occurs at a transition temperature $T_0$, where the Gibbs potential $G(P,T)$ becomes singular. The function $G$ is continuous, but some derivative becomes discontinuous across the transition point. When the first derivatives are discontinuous, we have a first-order transition, and when the first derivative is continuous but the second derivative is discontinuous, we have a second-order transition. The discontinuity of a thermodynamic function is a mathematical idealization in the macroscopic limit. For finite systems, no matter how large, thermodynamic functions are continuous with finite derivatives.

## 5.1.  First-Order Phase Transitions

The boiling and freezing of water are first-order transitions, in which the volume $V = (\partial G/\partial P)_T$ and the entropy $S = -(\partial G/\partial T)$ have different values in the two phases. The phase boundaries are shown in the $PT$ diagram in Fig. 5.1. Because of the difference in entropy, crossing the transition line at a fixed temperature necessitates the absorption or liberation of latent heat:

$$L_0 = T_0(s_2 - s_1) \tag{5.1}$$

where $s_i$ is the specific entropy of the $i$th phase. The specific entropy may mean entropy per particle, per mole, per unit mass or per unit volume. The specific heat at constant pressure, as a function

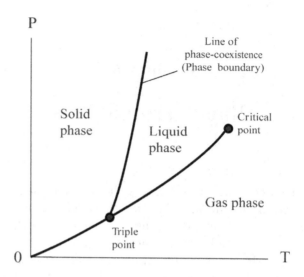

Fig. 5.1.   $PT$-diagram showing phase boundaries, or lines of phase-coexistence.

of temperature, contains a $\delta$-function at the transition point:

$$c_P = f(P, T) + L_0 \delta(T - T_0) \tag{5.2}$$

where $T_0$ and $L_0$ may be functions of $P$.

Because of the difference in specific volume of the coexisting phases, the isotherm in a $PV$-diagram exhibits a flat portion, as shown in Fig. 5.2 for a gas–liquid transition. At the point 1 the system is all liquid, at point 2 it is all gas, and in between, the system is a mixture of liquid and gas in states 1 and 2, respectively. The $PT$ and $PV$ diagrams are projections of the equation-of-state surface; shown in Fig. 5.3.

## 5.2.   Second-Order Phase Transitions

At the critical point of a gas–liquid transition, the two phases have equal density and specific entropy, and the phase transition becomes second order, associated with discontinuities in the second derivatives of the Gibbs potential. That is, specific heats are singular at the transition temperature $T_c$:

$$c_P \sim c_0 |t|^{-\alpha} \tag{5.3}$$

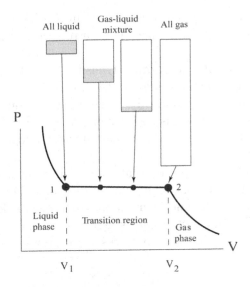

Fig. 5.2.   An isotherm in the gas–liquid transition region.

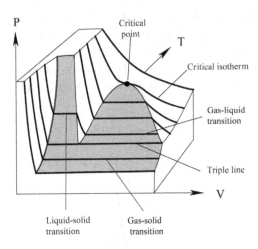

Fig. 5.3.   Equation of state surface.

where

$$t = \frac{T - T_c}{T_c} \tag{5.4}$$

with $\sim$ meaning "singular part is proportional to." The exponent $\alpha$ is the same whether we approach the critical point from above or below; but the proportional constant $c_0$ can be different, and may be zero on one side. Note that there is no $\delta$-function term because the latent heat is zero.

Other examples of second-order transitions are the ferromagnetic transition in magnetic materials, order–disorder transitions in metallic alloys, superconductivity and superfluid transitions. As an example, Fig. 5.4 shows the measured specific heat of liquid helium near the superfluid transition point. It has an nearly logarithmic singularity, with $\alpha \approx 0$.

As a general rule, the phase at lower temperature is in a more ordered state than the one at higher temperature. For example, in a ferromagnet, atomic magnetic moments are aligned below the critical temperature. The magnetization is as an "order parameter" in this sense. At the transition point it grows with a power-law behavior

$$M \sim m_0 |t|^\beta \tag{5.5}$$

where $m_0 = 0$ below the transition temperature and $\beta$ is another exponent like $\alpha$ in (5.3). This is illustrated in Fig. 5.5. The exponents $\alpha$ and $\beta$ are part of a set of critical exponents characteristic of the second-order phase transition. The power-law behaviors indicate an absence of a length scale at the critical point. This is borne out by

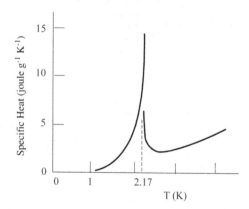

Fig. 5.4.   Experimental specific heat of liquid helium, showing a nearly logarithmic peak at the superfluid transition point. (Data from R.W. Hill and O.V. Lounasmaa (1957) *Phil. Mag.* **2**, 143.)

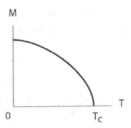

Fig. 5.5.   Magnetization in zero external field, as function of temperature.

the fact that the correlation length, which can be measured through scattering of light or neutrons, diverges:

$$\xi \sim k_0|t|^{-\nu} \tag{5.6}$$

This means that detailed structures of the system are not relevant at the critical point. Consequently, critical exponents have universality, in that they are shared by systems that may be very different in detailed structures.

## 5.3.   Van der Waals Equation of State

A very simple and instructive model for the gas–liquid phase transition is the van der Waals model. The potential energy $U(r)$ between two atoms as a function of their separation $r$ have the qualitative form shown in Fig. 5.6. To take the intermolecular interaction into account in a qualitative fashion, we separate the effects of the repulsive core and the attractive tail. The hard core excludes a certain

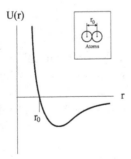

Fig. 5.6.   Interatomic potential.

volume around a molecule, so the other molecules have less room to move in. The effective volume is therefore smaller than the actual volume:

$$V_{\text{eff}} = V - b \tag{5.7}$$

where $V$ is the total volume of the system, and $b$ is the total excluded volume, of the order of $b \approx N\pi r_0^3/6$.

The pressure of the gas arises from molecules striking the wall of the container. Compared with the case of the ideal gas, a molecule in a real gas hits the wall with less kinetic energy, because it is being held back by the attraction of neighboring molecules. The reduction in the pressure is proportional to the number of pairs of interacting molecules near the wall, and thus to the density squared. Accordingly we have

$$P = P_{\text{kinetic}} - \frac{a}{V^2} \tag{5.8}$$

where $P_{\text{kinetic}}$ is the would-be pressure in the absence of attraction, and $a$ is a constant proportional to $N^2$. Van der Waals makes the assumption that, for 1 mole of gas,

$$V_{\text{eff}} P_{\text{kinetic}} = RT \tag{5.9}$$

where $R$ is the gas constant. This leads to the *van der Waals equation of state*

$$(V - b)\left(P + \frac{a}{V^2}\right) = RT \tag{5.10}$$

with isotherms as shown in Fig. 5.7. The pressure at fixed $T$ is a cubic polynomial in $V$:

$$(V - b)(PV^2 + a) = RTV^2$$
$$PV^3 - (bP + RT)V^2 + aV - ba = 0 \tag{5.11}$$

There is a region in which the polynomial has three real roots. As we increase $T$ these roots move closer together, and merge at $T = T_c$, the *critical point*. For $T > T_c$, one real root remains, while the other two become a complex-conjugate pair. We can find the critical parameters $P_c$, $V_c$, $T_c$, as follows. At the critical point, the equation of state must be of the form

$$(V - V_c)^3 = 0$$
$$V^3 - 3V_c V^2 + 3V_c^2 V - V_c^3 = 0 \tag{5.12}$$

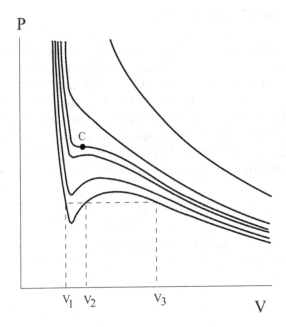

Fig. 5.7. Isotherms of the van der Waals equation of state.

Comparison with (5.11) yields

$$3V_c = b + \frac{RT_c}{P_c}, \quad 3V_c^2 = \frac{a}{P_c}, \quad V_c^3 = \frac{ba}{P_c} \tag{5.13}$$

They can be solved to give

$$RT_c = \frac{8a}{27b}, \quad P_c = \frac{a}{27b^2}, \quad V_c = 3b \tag{5.14}$$

Introducing the dimensional quantities

$$\bar{P} = \frac{P}{P_c}, \quad \bar{T} = \frac{T}{T_c}, \quad \bar{V} = \frac{V}{V_c} \tag{5.15}$$

we can rewrite the equation of state in the universal form

$$\left(\bar{V} - \frac{1}{3}\right)\left(\bar{P} + \frac{3}{\bar{V}^2}\right) = \frac{8}{3}\bar{T} \tag{5.16}$$

## 5.4.  Maxwell Construction

The van der Waals isotherm is a monotonic function of $V$ for $T > T_c$. Below $T_c$, however, there is a kink exhibiting negative compressibility. This is unphysical, and its origin can be traced to the implicit assumption that the density is always uniform. As we shall see, in thermal equilibrium the system undergoes a first-order phase transition, by breaking up into a mixture of phases of different densities.

According to the Maxwell relation $P = -(\partial A/\partial V)_T$, the free energy can obtained as the area under the isotherm:

$$A(V,T) = -\int_{\text{isotherm}} P\,dV \qquad (5.17)$$

Let us carry out the integration graphically, as indicated in Fig. 5.8. The volumes $V_1, V_2$ are defined by the double-tangent construction.

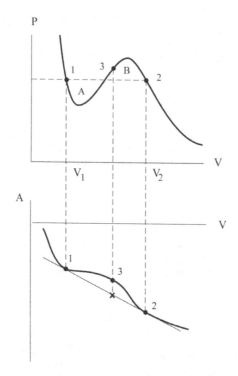

Fig. 5.8.  The Maxwell double-tangent construction for the free energy.

At any point along the tangent, such as X, the free energy is a linear combination of those at 1 and 2, and thus represents a mixture of two phases. This nonuniform state has the same $P$ and $T$ as the uniform state 3, but it has a lower free energy, as is obvious from the graphical construction. Therefore the phase-separated state is the equilibrium state. States 1 and 2 are defined by the conditions

$$\frac{\partial A}{\partial V_1} = \frac{\partial A}{\partial V_2} \quad \text{(Equal pressure)}$$

$$\frac{A_2 - A_1}{V_2 - V_1} = \frac{\partial A}{\partial V_1} \quad \text{(Common tangent)}$$

$$(5.18)$$

Thus,

$$-(A_2 - A_1) = -\frac{\partial A}{\partial V_1}(V_2 - V_1)$$

$$\int_{V_1}^{V_2} PdV = P_1(V_2 - V_1)$$

$$(5.19)$$

This means the areas $A$ and $B$ in Fig. 5.8 are equal.

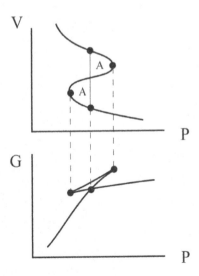

Fig. 5.9. The Maxwell construction corresponds to omitting the loop in the graph for the Gibbs potential $G$.

The Maxwell construction can also be made via the Gibbs potential. From the Maxwell relation $V = (\partial G/\partial P)_T$, we have

$$G(P,T) = \int_{\text{isotherm}} V dP \qquad (5.20)$$

The graphical integration is indicated in Fig. 5.9. In the graph for $G$, the loop is to be omitted because it lies higher in $G$. The graph is somewhat challenging to draw, because the two side arcs are concave downwards, while the top arc is concave upwards.

# Chapter 6

# Kinetics of Phase Transitions

## 6.1. Nucleation and Spinodal Decomposition

Along the van der Waals isotherm shown in Fig. 6.1, the system separates into a mixture of two phases between the points A and B, and follows the horizontal line. This describes the equilibrium situation. How does the system behave if it was initially prepared on the original isotherm? To answer this question, we examine an infinitesimal arc on the van der Waals isotherm, and repeat the argument used for the Maxwell construction, to see whether the free energy could be lowered by phase separation on a local scale, (with only infinitesimal change in the volume of the system). This is illustrated in Fig. 6.2.

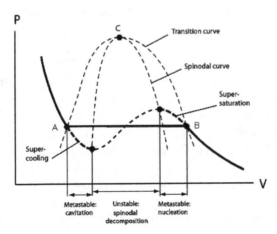

Fig. 6.1. The regions of supersaturation and supercooling correspond to metastable states, where phase transition is initiated via nucleation. Portions of the isotherm with negative compressibility $(-\partial P/\partial V < 0)$ represent unstable states that undergo spinodal decomposition — spontaneous phase separation.

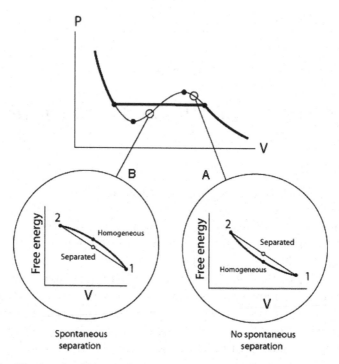

Fig. 6.2.   Portion A of the isotherm does not spontaneously separate into gas and liquid, but do so via nucleation, because the free energy has a positive curvature. Portion B separates spontaneously because the free energy has a negative curvature.

- In regions with positive compressibility ($\partial P/\partial V < 0$), the free energy has a positive curvature. Local phase separation does not occur, and the system is metastable. It is only metastable, because a global phase separation does lower the free energy, as the Maxwell construction shows. The global phase separation waits to be triggered by nucleation — the growth of a liquid droplet (or a gas bubble) created through random fluctuation. A computer simulation of nucleation in the freezing of water will be presented later.

- In regions with negative compressibility ($\partial P/\partial V > 0$), the free energy has a negative curvature. Local separation will occur spontaneously, and the system is unstable. This is called *spinodal decomposition*. The envelop of the unstable region is called the *spinodal curve*, as indicated in Fig. 6.1. The system will rapidly become a liquid–gas emulsion, and the liquid will gather into larger pools to minimize surface tension, on a longer time scale. The equation governing the decomposition, the Cahn–Hilliard equation, will be derived in the next chapter.

## 6.2. The Freezing of Water

Water is important for human life. It makes up more than 95% of a living cell, and it has a dominant presence in our environment. Yet, the properties of water are not well understood, because of the complexity of its molecular interactions. Water molecules have non-spherical shape, and attract each other via the hydrogen bond, which involves the sharing of a hydrogen atom by two oxygen atoms from different molecules. The bond length is about 2 A, as compared to the O–H separation of 0.9584 A in $H_2O$.

Because of hydrogen bonding, water molecules can form stable clusters of various sizes and shapes, some of which are shown in Fig. 6.3. The binding energy of a cluster has many-body contributions, and is not simply proportional to the number of bonds. The binding is strongest for pentagons and hexagons. The basic bond has a binding energy of 0.2 eV (5.5 kcal/mole), which is large compared with 0.026 eV, the thermal energy at room temperature. However, different water molecules compete for these bonds, so the bonding network is constantly changing. The average lifetime of a bond is of

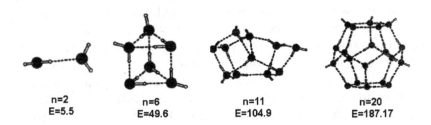

n=2
E=5.5

n=6
E=49.6

n=11
E=104.9

n=20
E=187.17

Fig. 6.3.   Water on a scale of $10^{-10}$ m. Clusters are formed by hydrogen bonding (dotted lines). The calculated binding energy $E$ is given in kcal/mole. [S. Maheshwary, N. Patel and N. Sathyamurthy (2001) *J. Phys. Chem.* **A105**, 10525–10537.]

Fig. 6.4.   Water on a scale of $10^2$ m.

the order of $10^{-12}$ s. The intricacy of the molecular network gives rise to amazing structures on a macroscopic scale, as illustrated in Fig. 6.4.

The phase transitions of water are not amenable to analytical studies. Computer studies are also difficult. But recently there has been a successful molecular dynamics simulation of the freezing of water, which throws light into the process of nucleation.[1] In this simulation, the coordinates of 512 water molecules were calculated at successive time steps, through numerical integrations of the Newtonian equations of motion, using known intermolecular potentials. The computation was carried out over a sufficiently large number of steps to cover the phase transition region. That this is feasible attests not only to the power of modern computers, but also the ingenuity of the investigators.

---

[1]M. Matsumoto, S. Saito and I. Ohmine (2002) *Nature* **416**, 409–412.

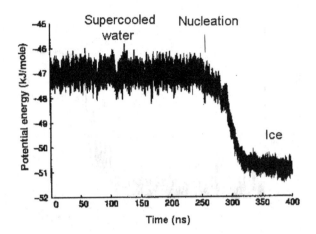

Fig. 6.5. Total potential energy as a function of time, after the system was quenched from a high temperature to below freezing point.

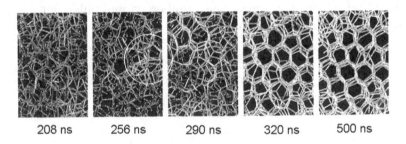

Fig. 6.6. Time development of the water network. Long-lasting bonds are shown bright. The encircled region contains a nucleous of ice. The last image exhibits the hexagonal crystalline structure of ice.

The calculations start by quenching the system from a high temperature to 230 K, below the freezing point, and then follow the dynamical time development. Figure 6.5 shows the total potential energy as a function of time given in ns. The water remains in a supercooled state for some time, waiting for a nucleus to be formed. (In this run the time taken was about 250 ns.) Thereafter, the nucleus rapidly grows until the entry system is frozen. The formation time for a nucleus varies in different runs, exhibiting a Poisson distribution. This indicates that nucleation occurred at random. Figure 6.6

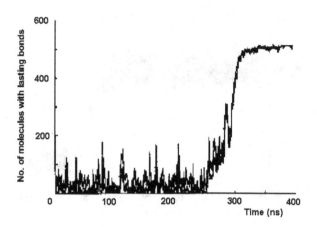

Fig. 6.7.   Number of molecules with long-lasting ($> 2\,\mathrm{ns}$) bonds, as a function of time. Before reaching the plateau, the number fluctuation was $1/f$ noise.

shows the structure of the hydrogen bonds at different times after quenching. Figure 6.7 shows the number of molecules with long-lasting hydrogen bonds, as a function of time.

# Chapter 7

# The Order Parameter

## 7.1. Ginsburg–Landau Theory

As we have noted, in a phase transition, a system changes from a less ordered state to a more ordered one, or vice versa. As an example of a measure of order, we mentioned the magnetization of a ferromagnet, which is nonzero only below the transition temperature, when the atomic magnetic moments are aligned.

In the Ginsburg–Landau theory, the state of the system is described by a local *order parameter* $\phi(x)$, a scalar field modeled after the magnetization density. The statistical properties of the system is obtained through an ensemble of functions $\phi(x)$, weighted by the Boltzmann factor

$$e^{-\beta E[\phi]} \tag{7.1}$$

where the functional $E[\phi]$ has the form

$$E[\phi] = \int dx \left[ \frac{\varepsilon^2}{2} \left| \nabla \phi(x) \right|^2 + W(\phi(x)) - h(x)\phi(x) \right] \tag{7.2}$$

which is called the *Landau free energy*.[1] It contains a kinetic term $|\nabla \phi|^2$, a potential $W(\phi)$ and external field $h(x)$. The constant $\varepsilon$ is a "stiffness coefficient." The kinetic term imposes an energy cost for a

---

[1]This could be called the "van der Waals free energy," for it was first proposed in J.D. van der Waals (1893) The thermodynamic theory of capillarity flow under the hypothesis of continuous variation in density, *Verhandelingen der Koninklijke Nederlansche Akademie van Wetenchappen te Amsterdam* **1**, 1–56.

gradient of the field $\phi$, and drives the system towards uniformity. In general, $x$ is a position in $D$ dimensions.

The partition function in the presence of an external field $h(x)$ is given by

$$Q[h] = \int D\phi e^{-\beta E[\phi]} \tag{7.3}$$

and the ensemble average of $O$ is given by

$$\langle O \rangle = \frac{\int D\phi O e^{-\beta E[\phi]}}{\int D\phi e^{-\beta E[\phi]}} \tag{7.4}$$

where $\int D\phi$ denotes functional integration over all functional forms of the field $\phi$. The implicit assumption is that all variables of the system other than the order parameter have been integrated out, and $E[\phi]$ embodies the results of these integrations. This is why $E[\phi]$ is called a free energy. We thus expect the potential $W$ to be dependent on the temperature.

We expect the functional integrals to be dominated by the maximum values of their integrands, and the latter correspond to the $\phi$ that minimizes the Landau free energy. This $\phi$ gives the thermodynamic properties of the system, and functions in its neighborhood represent thermal fluctuations.

## 7.2.  Second-Order Phase Transition

We consider $h = 0$. To model a second-order phase transition, choose

$$W(\phi) = r_0\phi^2 + u_0\phi^4$$
$$r_0 = bt \tag{7.5}$$

where $b$ is a positive real constant, and

$$t = \frac{T - T_c}{T_c} \tag{7.6}$$

The parameter $u_0 > 0$ is independent of $t$, but $r_0$ changes sign at the transition point. The potential has the shape shown in Fig. 7.1.

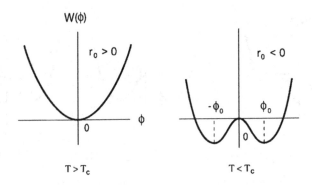

Fig. 7.1.   Modeling a second-order phase transition.

The field that minimizes the Landau free energy is spatially constant, with the value

$$\phi_0 = \begin{cases} 0 & (T > T_c) \\ \pm\sqrt{b(T - T_c)/u_0} & (T < T_c) \end{cases} \tag{7.7}$$

We choose one of the $\pm$ signs. (This is known as spontaneous symmetry breaking.)

In an approximation known as *mean-field theory*, we neglect the fluctuations about $\phi_0$. Then the magnetization is $M = V\phi_0$, where $V$ is the volume of the system. This gives a critical exponent $\beta = 1/2$. On including the fluctuations, corrections are made to this value.

We see that discontinuous behavior, such as a phase transition, can arise from continuous changes in continuous functions. This insight is the basis of the so-called "catastrophe theory."

## 7.3.   First-Order Phase Transition

We can describe a first-order transition by assuming a form of $W$ such that, as the temperature varies, the function assumes a sequence of shapes as shown in Fig. 7.2. We interpret $\phi$ as the particle density. The two minima at $\phi_1, \phi_2$ represent respectively the low-density and high-density phases of the system. Just below the transition temperature, $\phi_1$ corresponds to the stable phase. At the transition point,

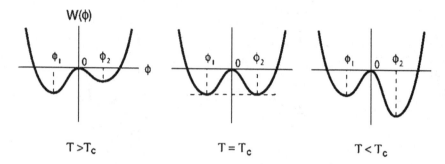

Fig. 7.2.   Modeling a first-order phase transition.

both $\phi_1$ and $\phi_2$ are stable, and just above the transition point $\phi_2$ becomes the stable phase.

The present theory is phenomenological, and we can choose $W$ in any convenient way to suit our purpose. It is interesting to note, however, that $W$ can be chosen to correspond to the van der Waals equation of state. For uniform $\phi$ in the absence of external field, the Landau free energy is

$$E(\phi) = VW(\phi) \tag{7.8}$$

where the total volume $V$ is a fixed parameter. The relevant variable is the specific volume $1/\phi$. Thus, the pressure is given by the Maxwell relation

$$P = -\frac{1}{V}\frac{\partial E(\phi)}{\partial(1/\phi)} = \phi^2 W'(\phi) \tag{7.9}$$

where $W'(\phi) = \partial W/\partial\phi$. Equating this to the van der Waals equation of state leads to

$$W'(\phi) = \frac{RT}{\phi(1 - b\phi)} - a$$
$$W(\phi) = RT\ln\frac{\phi}{1 - b\phi} - a\phi + c \tag{7.10}$$

where $c$ is a constant.

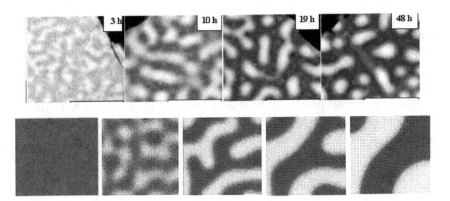

Fig. 7.3. The upper panel shows X-ray microscopy images of the time development of spinodal decomposition in a thin film of polymer blend [Courtesy Professor H. Ade, NCSU]. The lower panel shows the phase separation of a 2D binary alloy, from computer simulations of the Cahn–Hilliard equation (www.lam.uni-bonn.de/grape/examples/cg/ch/html).

## 7.4. Cahn–Hilliard Equation

We now describe the dynamics of the system. The free energy per particle at position $\mathbf{r}$ is proportional to the functional derivative

$$\frac{\delta E[\phi]}{\delta \phi(\mathbf{r})} = -\varepsilon^2 \nabla^2 \phi + W'(\phi) \tag{7.11}$$

A nonuniformity in this quantity will induce a diffusion current proportional to its gradient:

$$\mathbf{j} = -\nabla \left[ -\varepsilon^2 \nabla^2 \phi + W'(\phi) \right] \tag{7.12}$$

where the proportionality constant has been absorbed into $\varepsilon$ and $W$. From the continuity equation

$$\frac{\partial \phi}{\partial t} + \nabla \cdot \mathbf{j} = \mathbf{0} \tag{7.13}$$

we obtain the dynamical equation

$$\frac{\partial \phi}{\partial t} = -\nabla^2 \left[ \varepsilon^2 \nabla^2 \phi - W'(\phi) \right] \tag{7.14}$$

which is known as the Cahn–Hilliard equation.[2]

---

[2] J.W. Cahn and J.E. Hilliard (1958) *J. Chem. Phys.* **28**, 258–267. See C.P. Grant (1993) *Comm. Partial Diff. Equ.* **18**, 453–490, for a rigorous treatment of the 1D equation.

The Cahn–Hilliard equation is a nonlinear dissipative equation, and its solution approaches a limiting form at large times. In this respect, it is similar to the Navier–Stokes equation in hydrodynamics. Computer simulations of this equation, with simple polynomial potentials, have been successful in describing the dynamics of spinodal decomposition. This is shown in Fig. 7.3 for a hypothetical substance. Experimental images from a real system are included for comparison.

# Chapter 8

# Correlation Function

## 8.1. Correlation Length

The correlation function for a field $\phi(x)$ is defined as

$$G(x, y) = \langle \phi(x)\, \phi(y) \rangle - \langle \phi(x) \rangle \langle \phi(y) \rangle \tag{8.1}$$

If the joint average $\langle \phi(x)\phi(y) \rangle$ is the same as the product of the individual averages, then by definition there is no correlation between the values of the field at $x$ and $y$, and $G(x, y) = 0$. The correlation function can be directly measured through X-ray or neutron scattering.[1] Generally, $G(x, y)$ decreases with increasing distance $|x - y|$ exponentially, for large distances:

$$G(x, y) \sim e^{-|x-y|/\xi} \tag{8.2}$$

The characteristic length $\xi$ is called the *correlation length*. We shall show that it diverges at the critical point.

## 8.2. Large-Distance Correlations

Let

$$g(x) = G(x, 0) \tag{8.3}$$

which is the relative probability of finding a particle at $x$, knowing that there is one at the origin. We calculate this approximately by finding the mean-field in the presence of a point source at the origin:

$$h(x) = h_0 \delta^D(x) \tag{8.4}$$

---

[1] J. Als-Nielsen and D. McMorrow, *Elements of X-Ray Physics* (Wiley, NY, 2001).

The Landau free energy is then

$$E[\phi] = \int d^D x \left[ \frac{\varepsilon}{2} |\nabla \phi(x)|^2 + r_0 \phi^2(x) + u_0 \phi^4(x) - h_0 \phi(x) \delta^D \right]$$

$$= \int d^D x \left[ -\frac{\varepsilon}{2} \phi \nabla^2 \phi + r_0 \phi^2 + u_0 \phi^4 - h_0 \phi \delta^D \right] \tag{8.5}$$

where $r_0 = bt$, and the term $-\nabla^2 \phi$ on the second line is arrived at by performing a partial integration. In the mean-field approximation, we take $g(x)$ to the field that minimizes $E[\phi]$. For a small variation about the mean field

$$\phi(x) = g(x) + \delta \phi(x) \tag{8.6}$$

the first-order variation of $E[\phi]$ should vanish:

$$0 = \delta E[\phi] = \int d^D x [-\varepsilon \nabla^2 g + 2 r_0^2 g + 4 u_0 g^3 - h_0 \delta^D] \delta \phi \tag{8.7}$$

Since $\delta \phi$ is arbitrary, we obtain the equation

$$-\varepsilon \nabla^2 g(x) + 2 r_0 g(x) + 4 u_0 g^3(x) = h_0 \delta^D(x) \tag{8.8}$$

This is the *nonlinear Schrödinger equation* with a unit source. It is the simplest nonlinear generalization of the Schrödinger equation, and occurs in such diverse fields as plasma physics, quantum optics, superfluidity, and the theory of elementary particles.

We now neglect the nonlinear $g^3$ term, and this should be a good approximation when $g(x) \to 0$, since then $g^3 \ll g^2$, unless $r_0 = 0$. Thus we have the linear equation

$$-\varepsilon \nabla^2 g(x) + 2 r_0 g(x) = h_0 \delta^D(x) \tag{8.9}$$

Taking the Fourier transform of both sides, we obtain

$$(\varepsilon k^2 + 2 r_0) \tilde{g}(k) = h_0 \tag{8.10}$$

where $\tilde{g}(k)$ is the Fourier transform of $g(x)$:

$$\tilde{g}(k) = \int d^D x e^{-ik \cdot x} g(x) \tag{8.11}$$

The solution is

$$\tilde{g}(k) = \frac{h_0}{\varepsilon k^2 + 2 r_0} \tag{8.12}$$

and the inverse transform gives

$$g(x) = h_0 \int \frac{d^D k}{(2\pi)^D} \frac{e^{ik \cdot x}}{\varepsilon k^2 + 2r_0} \tag{8.13}$$

For large $|x|$ for $D > 2$ we have the asymptotic behavior

$$g(x) \approx C_0 |x|^{2-D} e^{-|x|/\xi} \tag{8.14}$$

where $C_0$ is a constant, and

$$\xi = (2r_0)^{-1/2} = (2b)^{-1/2} t^{-1/2} \quad (t > 0) \tag{8.15}$$

Thus, the correlation length diverges as $t \to 0$, with critical exponent $1/2$, in our approximation.

At precisely the critical point $t = 0$, (8.14) gives, in our approximation, $g \propto |x|^{2-D}$. But, as noted before, the linear approximation may not be valid at $t = 0$. Indeed, the correct critical behavior is

$$g(x) \propto |x|^{2-D-\eta}$$
$$\tilde{g}(k) \propto k^{-2+\eta} \tag{8.16}$$

where $\eta$ is a critical exponent. The dimension of space seems to have changed from $D$ to $D + \eta$, which is called the "anomalous dimension."

Actually, we need not have $t = 0$ to have power-law behavior. As long as $|x| \ll \xi$, the correlation length is effectively infinite, and we will have (8.16).

## 8.3. Universality Classes

The correlation length $\xi$ measures the distance within which values of the field are correlated. We cannot resolve spatial structures smaller than $\xi$, because the field organizes itself into uniform blocks of approximately that size. As we approach the critical point, $\xi$ increases, and we lose resolution. At the critical point, when $\xi$ diverges, we cannot see any details at all. Only global properties, such as the dimension of space, or the number of degrees of freedom, distinguish one system from another. That is why systems at the critical point fall into universality classes, which share the same set of critical exponents.

## 8.4.  Compactness Index

For a distribution of $N$ points set up by a given rule, the characteristic length $R$ of the system depends on $N$, as $N \to \infty$, through a power law

$$R = aN^\nu \tag{8.17}$$

where $a$ is a length scale. The exponent $\nu$ is an indication of the compactness of the system. For example:

- For ordinary matter $\nu = 1/3$, since it "saturates," i.e. the density $N/R^3$ is constant.
- For Brownian motion $\nu = 1/2$, since the average end-to-end distance for $N$ steps is proportional to $\sqrt{N}$.

The Fourier transform of the correlation function at high wave numbers has the behavior

$$\tilde{g}(k) \sim k^{-1/\nu} \quad (kR \gg 1) \tag{8.18}$$

Thus the index $\nu$ can be measured through scattering experiments.


## 8.5.  Scaling Properties

The property (8.18) can be shown by considering the scaling properties of the correlation function.[2] The Landau free energy $E[\phi]$ is of dimension energy. Let us choose the constant $\varepsilon$ to be of dimension energy. Then the field $\phi(x)$ has dimension $L^{-D/2}$, the correlation function $g(x)$ has dimension $L^{-D}$, and $\tilde{g}(k)$ is dimensionless. Thus, the latter can depend on $k$ only through $ka$, and we can represent it in the form

$$\tilde{g}(k) = F(ka, N) \tag{8.19}$$

where $F$ is a dimensionless function.

---

[2]P.G. de Gennes, *Scaling Concepts in Polymer Physics* (Cornell University Press, Ithaca, 1979).

The relation $R = aN^\nu$ is invariant under the scale transformation

$$a \to a\lambda^\nu$$
$$N \to \frac{N}{\lambda} \tag{8.20}$$

Now, $\tilde{g}(k)$ should scale like $N$, since it gives the scattering cross-section from $N$ scatterers. Thus it should transform according to

$$\tilde{g}(k) \to \frac{1}{\lambda}\tilde{g}(k)$$

Using (8.19), we have

$$F(ka, N) = \frac{1}{\lambda} F\left(ka\lambda^\nu, \frac{N}{\lambda}\right) \tag{8.21}$$

for arbitrary $\lambda$. This is satisfied by the functional form

$$F(ka, N) = Nf(kaN^\nu) \tag{8.22}$$

Thus

$$\tilde{g}(k) = Nf(kaN^\nu) = Nf(kR) \tag{8.23}$$

The main result of the scaling argument is that $\tilde{g}(k)/N$ depends on $N$ only through $kR$.

Now we make the following arguments:

- $g(x)$ should be independent of $N$ for $|x| \ll R$, because it should depend on local properties only.
- Hence $\tilde{g}(k)$ should be independent of $N$ for $kR \gg 1$.

Assuming the power behavior $f(kR) \sim (kR)^p$ for large $kR$, we have:

$$\tilde{g}(k) \sim N(kR)^p \sim N(kaN^\nu)^p \tag{8.24}$$

This can be independent of $N$ only if $\nu p + 1 = 0$, hence $p = -1/\nu$, and we obtain the result (8.18).

Since the result depends on the small-distance behavior of $g(x)$, the correlation length is effectively infinite, and $\nu$ is a critical exponent. Comparison with (8.16) shows it is related to the anomalous dimension through

$$\nu = \frac{1}{2 - \eta} \tag{8.25}$$

# Chapter 9

# Stochastic Processes

## 9.1. Brownian Motion

Macroscopic phenomena rarely give any hint of thermal fluctuations. Indeed, the smallness of these fluctuations underlies the effectiveness of thermodynamics. In Brownian motion, however, we can actually see such fluctuations under a microscope. This is because the micron-sized Brownian particles, suspended in solution, are sufficiently small so impacts from solvent atoms do not completely cancel out.

Figure 9.1 reproduces some of the pioneering observations made by Perrin in 1909. The positions of a Brownian particle were made at intervals of 30 s, during which time it experiences the order of $10^{21}$ atomic impacts. Thus, in any observed interval, the final state of a Brownian particle is statistically independent of the initial state, because even an infinitesimal change in the initial condition would lead to a finite difference in the final state. This extreme sensitivity on the initial condition defines *randomness*. The fit of Perrin's data to the diffusion law, as shown in Fig. 9.1(c), verifies the random nature of Brownian motion.

If we reduce the observation time interval by factors 10, 100, 1,000, ..., the observed motion would still look the same. There is self-similarity under a change of time scale, and this will be so until the observation time becomes the order of atomic time scale. Not only is the atomic time scale a long way off, but to make observations on that time scale would require a higher technology.

Einstein was the first to offer a theory of Brownian motion in 1905. His great insight was that each Brownian step must be regarded

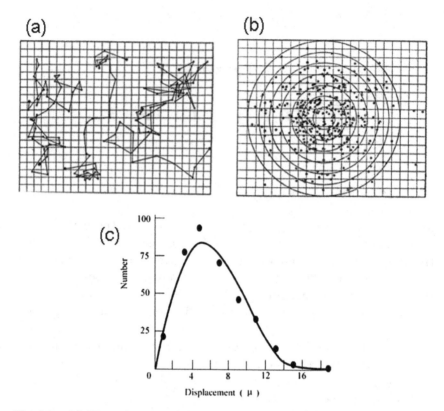

Fig. 9.1.   (a) 2D projections of Brownian paths recorded by Perrin. (b) Spatial distribtution of Brownian particles, with all initial positions translated to a common center. (c) Graph of the distance distribution. The solid curve is calculated from the diffusion law.

as the resultant of a large number of random steps, and must therefore be treated in a statistical manner.

A variable that is the sum of a large number of random contributions is now called a *stochastic variable*. The time development of such a variable is called a *stochastic process*. It cannot be described by a function of time in the ordinary sense, for it must be represented by a sequence of probable values. Instead of a single history, we have to consider an ensemble of histories. The simplest stochastic process is the one-dimensional random walk.

## 9.2. Random Walk

A 1D random walk refers to a particle on a linear discrete lattice. At each time step, the particle has equal probability of moving forward or backward. The total number of possibilities for making $n$ steps is $2^n$, for there are two possibilities for each step. The probability $W(k, n)$ that, after making $n$ steps, the particle finds itself $k$ steps from the starting point is thus given by

$$W(k, n) = \frac{\text{No. of walks with the same } k}{2^n} \tag{9.1}$$

Let $f$ and $b$ denote respectively the number of forward and backward steps made. Clearly

$$\begin{aligned} f + b &= n \\ f - b &= k \end{aligned} \tag{9.2}$$

Solving these, we obtain

$$f = \frac{n + k}{2} \tag{9.3}$$
$$b = \frac{n - k}{2}$$

The number of walks with the same $k$ is equal to the number of walks with the same $f$. The latter can be chosen in $\binom{n}{f}$ ways. Thus

$$W(k, n) = \binom{n}{f} \frac{1}{2^n} = \frac{n!}{f!(n - f)!} \frac{1}{2^n} \tag{9.4}$$

or

$$W(k, n) = \frac{2^{-n} n!}{\left(\frac{n+k}{2}\right)! \left(\frac{n-k}{2}\right)!} \tag{9.5}$$

The is called a *binomial distribution* in $k$.

For $n \gg k \gg 1$, we approximate the factorials using the Stirling approximation, and expand the result to lowest order in $k/n$. The result is[1]

$$W(k, n) \approx \sqrt{\frac{2}{\pi n}} e^{-k^2/2n} \tag{9.6}$$

This is called a *Gaussian distribution in k.*

## 9.3.  Diffusion

We now translate this result to a physical situation. The distance covered is $x = kx_0$, where $x_0$ is a unit of length. The time taken is $t = nt_0$, where $t_0$ is the duration of a step in seconds. For large $k$, the distance is the resultant of a large number of random steps, hence a stochastic variable. The probability density of finding the particle at $x$ is given by the diffusion law

$$W(x, t) = \frac{1}{\sqrt{4\pi Dt}} e^{-x^2/4Dt} \tag{9.7}$$

where $D = x_0^2/2t_0$ is the diffusion constant. The normalization is such that

$$\int_{-\infty}^{\infty} dx W(x, t) = 1$$
$$W(x, t) \xrightarrow[t\to 0]{} \delta(x) \tag{9.8}$$

The mean-square average of the position is just the variance of the distribution

$$\langle x^2 \rangle = \frac{\int_{-\infty}^{\infty} dx\, x^2 W(x, t)}{\int_{-\infty}^{\infty} dx W(x, t)} = 2Dt \tag{9.9}$$

---

[1]To obtain the prefactor in front of the exponential, one needs a more accurate form of the Stirling approximation

$$\ln n! \approx n \ln n - n + \ln \sqrt{2\pi n}$$

However, the prefactor can be obtained simply by noting the requirement

$$\int_{-\infty}^{\infty} dk\, W(k, n) = 1$$

since $W(k, n)$ is a probability.

which increases linearly with time. That is, after a time $t$, the probable distance of the particle from the starting point is $\sqrt{4Dt}$.

The probability distribution in 3D is just a product of the 1D probability densities:

$$W(\mathbf{r}, t) = \frac{1}{(4\pi Dt)^{3/2}} e^{-r^2/4Dt} \tag{9.10}$$

To project the 3D distribution onto the $x$–$y$ plane, use cylindrical coordinates $\rho, \theta, z$, where $\rho$ is the distance from the origin in the $x$–$y$ plane, and calculate

$$\rho d\rho \int_0^{2\pi} d\theta \int_{-\infty}^{\infty} dz\, W(\mathbf{r}, t) = \frac{2\pi \rho d\rho}{(4\pi Dt)^{3/2}} \int_{-\infty}^{\infty} dz\, e^{-(\rho^2 + z^2)/4Dt}$$

$$= \frac{\rho d\rho}{2Dt} e^{-\rho^2/4Dt} \tag{9.11}$$

The projected distribution is therefore

$$P(\rho, t) = \frac{\rho}{2Dt} e^{-\rho^2/4Dt} \tag{9.12}$$

The theoretical curve in Fig. 9.1(c) is proportional to the above with $\sqrt{4Dt} = 7.16\,\mu$.

## 9.4.  Central Limit Theorem

Our result shows that:

> *The sum of a large number of stochastic variables*
>
> *obeys a Gaussian distribution.* $\qquad$ (9.13)

This is known as the *central limit theorem*. It explains why the "bell curve" of the Gaussian distribution is so prevalent, from error analysis in laboratory experiments to age distribution in a population.

## 9.5.  Diffusion Equation

We can derive the diffusion law from a different physical point of view. Suppose that, at an initial time, there were $N$ particles in a medium, placed at the origin in 3D space at time $t = 0$, so that the density is $N\delta^3(\mathbf{r})$. What is the density $n(\mathbf{r}, t)$ at time $t$?

Since the number of particles is conserved, the current density $\mathbf{j}(\mathbf{r}, t)$ associated with a flow of the particles satisfies the continuity equation

$$\frac{\partial n}{\partial t} + \nabla \cdot \mathbf{j} = 0 \tag{9.14}$$

Now assume the phenomenological relation

$$\mathbf{j} = -D\nabla n \tag{9.15}$$

which defines the diffusion coefficient $D$. Combining the two equations, we obtain the diffusion equation

$$\frac{\partial n}{\partial t} = D\nabla^2 n \tag{9.16}$$

with normalization

$$\int d^3 r\, n(\mathbf{r}, t) = N \tag{9.17}$$

The solution is

$$n(\mathbf{r}, t) = \frac{N}{(4\pi D t)^{3/2}} e^{-r^2/4Dt} \tag{9.18}$$

Setting $N = 1$ reduces the above to the probability density $W(\mathbf{r}, t)$.

# Chapter 10

# Langevin Equation

## 10.1. The Equation

The motion of a Brownian particle may be described by the Langevin equation:

$$m\frac{dv}{dt} + \gamma v = F(t) \tag{10.1}$$

where $v = dx/dt$ is the velocity and $m$ is the mass of the particle. The force acting on the particle by the environment is split into a friction $-\gamma v$, and a random force $F(t)$. Interactions between different Brownian particles are neglected.

If $F$ were an external force that drags the particle in the medium, then the terminal velocity would be given by $v = F/\gamma$. Thus, $\gamma^{-1}$ is the mobility of the particle.

To treat $F$ as a random force, we must consider an ensemble of systems, and define the random force through its ensemble averages:

$$\langle F(t) \rangle = 0$$
$$\langle F(t_1)F(t_2) \rangle = c_0\delta(t_1 - t_2) \tag{10.2}$$

where $c_0$ is a constant, and the correlation time of the force is taken to be zero. That is to say, the correlation time is assumed to be much shorter than any other time scale in the problem. This is a reasonable approximation, for the atomic collision time is of order $10^{-20}$ s. The random force so defined corresponds to "white noise."

Because of the ensemble interpretation, the velocity $v$ and position $x$ are stochastic variables. According to the central limit theorem, they should both have Gaussian distributions in the steady state. The variances of these distributions are independently known, for the $x$ distribution should obey the diffusion law, and the velocity distribution should be Maxwell–Boltzmann. Thus we must have

$$\langle x^2 \rangle = 2Dt$$
$$m\langle v^2 \rangle = kT \tag{10.3}$$

where $D$ is the diffusion coefficient, $t$ the time, and $T$ the absolute temperature. It has been assumed that $x = 0$ at $t = 0$.

By calculating these variances via the Langevin equation, we can relate the parameters $c_0$ and $\gamma$ to physical properties. We shall show

$$\frac{c_0}{2\gamma} = kT$$
$$\gamma = \frac{kT}{D} \tag{10.4}$$

The first relation is known as the fluctuation–dissipation theorem, and the second is Einstein's relation.

## 10.2.  Solution

To solve the Langevin equation, we take Fourier transforms:

$$v(t) = \int_{-\infty}^{\infty} \frac{d\omega}{2\pi} e^{-i\omega t} u(\omega)$$
$$F(t) = \int_{-\infty}^{\infty} \frac{d\omega}{2\pi} e^{-i\omega t} f(\omega) \tag{10.5}$$

The Fourier transform of the random force $f(\omega)$ satisfies

$$\langle f(\omega) \rangle = 0$$
$$\langle f(\omega) f(\omega') \rangle = 2\pi c_0 \delta(\omega + \omega') \tag{10.6}$$

The transform of the Langevin equation reads

$$-im\omega u + \gamma u = f \tag{10.7}$$

Solving for $u$, we obtain

$$u(\omega) = \frac{f(\omega)}{\gamma - im\omega} \tag{10.8}$$

## 10.3.   Fluctuation–Dissipation Theorem

We now calculate $\langle v^2 \rangle$. The velocity correlation function in Fourier space is

$$\langle u(\omega)u(\omega') \rangle = \frac{2\pi c_0 \delta(\omega + \omega')}{\gamma^2 + m^2\omega^2} \tag{10.9}$$

The inverse Fourier transform gives

$$\langle v(t)v(t') \rangle = \int_{-\infty}^{\infty} \frac{d\omega}{2\pi} e^{-i\omega t} \int_{-\infty}^{\infty} \frac{d\omega'}{2\pi} e^{-i\omega' t'} \frac{2\pi c_0 \delta(\omega + \omega')}{\gamma^2 + m^2\omega^2}$$

$$= c_0 \int_{-\infty}^{\infty} \frac{d\omega}{2\pi} \frac{e^{-i\omega(t-t')}}{\gamma^2 + m^2\omega^2} = c_0 \int_{0}^{\infty} \frac{d\omega}{\pi} \frac{e^{-i\omega(t-t')}}{\gamma^2 + m^2\omega^2} \tag{10.10}$$

Taking $t = t'$, we obtain

$$\frac{m}{2} \langle v^2 \rangle = mc_0 \int_{0}^{\infty} \frac{d\omega}{2\pi} \frac{1}{\gamma^2 + m^2\omega^2} = \frac{c_0}{4\gamma} \tag{10.11}$$

In thermal equilibrium this should be equal to $kT/2$ by the equipartition of energy. Hence

$$\frac{c_0}{2\gamma} = kT \tag{10.12}$$

This is called the *fluctuation–dissipation theorem*. It relates two aspects of the external force: random force strength $c_0$ (fluctuation), and the damping constant $\gamma$ (dissipation).

## 10.4.   Power Spectrum and Correlation

In (10.11) we can identify the integrand as the energy per unit frequency or the *power spectrum*:

$$S(\omega) = \frac{mc_0}{\gamma^2 + m^2\omega^2} \tag{10.13}$$

Using the fluctuation–dissipation theorem, we can write

$$S(\omega) = \frac{2m\gamma kT}{\gamma^2 + m^2\omega^2} \tag{10.14}$$

In the limit of small dissipation we have

$$S(\omega) \xrightarrow[\gamma \to 0]{} 2\pi kT\, \delta(\omega) \qquad (10.15)$$

Now we rewrite (10.10) in the form

$$\frac{m}{2}\langle v(t)v(0)\rangle = \int_0^\infty \frac{d\omega}{2\pi} e^{-i\omega t} S(\omega) \qquad (10.16)$$

The inversion of this formula gives

$$S(\omega) = m \int_0^\infty dt\, e^{-i\omega t} \langle v(t)v(0)\rangle \qquad (10.17)$$

The power spectrum and the velocity correlation functions are Fourier transforms of each other. This statement is sometimes called the "Wiener–Kintchine theorem."

## 10.5.  Causality

The calculation of $\langle x^2 \rangle$ is more intricate. In this section, we set $m = 1$ for simplicity.

Let the Fourier transform of $x(t)$ be $z(\omega)$. Since $dx/dt = v$, we have $-i\omega z(\omega) = u(\omega)$. Thus

$$z(\omega) = \frac{i}{\omega}\frac{f(\omega)}{\gamma - i\omega} \qquad (10.18)$$

and

$$x(t) = \int_{-\infty}^\infty \frac{d\omega}{2\pi} e^{-i\omega t} \frac{i}{\omega}\frac{f(\omega)}{\gamma - i\omega} \qquad (10.19)$$

We have to deform the integration contour to circumvent the pole at the origin. The requirement of causality dictates that we must detour the path above the pole, as we now show.

Consider the correlation function

$$\langle z(\omega)f(\omega')\rangle = \frac{i}{\omega}\langle u(\omega)f(\omega')\rangle = \frac{i2\pi c_0 \delta(\omega + \omega')}{\omega(\gamma - i\omega)} \qquad (10.20)$$

The inverse transform gives

$$\langle x(t)F(t')\rangle = \int_{-\infty}^{\infty} \frac{d\omega}{2\pi} e^{-i\omega t} \int_{-\infty}^{\infty} \frac{d\omega'}{2\pi} e^{-i\omega' t'} \frac{i2\pi c_0 \delta(\omega + \omega')}{\omega(\gamma - i\omega)}$$

$$= -c_0 \int_{-\infty}^{\infty} \frac{d\omega}{2\pi} e^{-i\omega(t-t')} \frac{1}{\omega(\omega + i\gamma)} \tag{10.21}$$

Again, this is ambiguous because of the pole at the origin. We must treat the pole in such a manner as to make

$$\langle x(t)F(t')\rangle = 0 \quad (t < t') \tag{10.22}$$

which is required by causality, i.e. the position cannot respond to a force applied in the future. This can be achieved by deforming the path above $\omega = 0$. Equivalently, we displace the pole to the lower half plane through the replacement

$$\frac{1}{\omega} \to \frac{1}{\omega + i\epsilon} \quad (\epsilon \to 0^+) \tag{10.23}$$

We then obtain, through contour integration,

$$\langle x(t)F(t')\rangle = \begin{cases} \frac{c_0}{\gamma}\left[1 - e^{-\gamma(t-t')}\right] & (t > t') \\ 0 & (t < t') \end{cases} \tag{10.24}$$

Note that there is no correlation at equal time:

$$\langle x(t)F(t)\rangle = 0 \tag{10.25}$$

The position coordinate is now well defined:

$$x(t) = \int_{-\infty}^{\infty} \frac{d\omega}{2\pi} e^{-i\omega t} \frac{i}{\omega + i\epsilon} \frac{f(\omega)}{\gamma - i\omega} \quad (\epsilon \to 0^+) \tag{10.26}$$

We can now calculate the mean-square displacement, resulting in

$$\langle [x(t) - x(0)]^2 \rangle = \frac{2c_0}{\pi} \int_{-\infty}^{\infty} d\omega \frac{\sin^2(\omega t/2)}{(\omega^2 + \epsilon^2)(\omega^2 + \gamma^2)} \tag{10.27}$$

where we can safely set $\epsilon = 0$. In the limit $t \to \infty$, we can use the formula

$$\frac{\sin^2(wt)}{w^2} \xrightarrow[t\to\infty]{} \pi t \, \delta(w) \tag{10.28}$$

Thus

$$\langle [x(t) - x(0)]^2 \rangle \xrightarrow[t \to \infty]{} \frac{c_0 t}{\gamma^2} \tag{10.29}$$

Since the variance of the $x$ distribution should be $2Dt$, we have

$$c_0 = 2\gamma^2 D \tag{10.30}$$

Combined with the fluctuation–dissipation theorem $c_0/2\gamma = kT$, we obtain

$$\gamma = \frac{kT}{D} \tag{10.31}$$

This is Einstein's relation.

## 10.6.   Energy Balance

The average kinetic energy is

$$K = \frac{m}{2} \langle v^2 \rangle \tag{10.32}$$

Multiply both sides of the Langevin equation by $v$, we have

$$\frac{m}{2} \frac{dv^2}{dt} + \gamma v^2 = vF \tag{10.33}$$

Taking the average yields

$$\frac{dK}{dt} = \langle vF \rangle - \frac{2\gamma}{m} K \tag{10.34}$$

where

$$\langle vF \rangle = \text{Rate of work done on the system}$$
$$\frac{2\gamma}{m} K = \text{Rate of energy dissipation} \tag{10.35}$$

In the steady state, a dynamic equilibrium is maintained by balancing energy input and dissipation: $2\gamma K = m \langle vF \rangle$.

We can calculate $\langle vF \rangle$ as follows:

$$
\begin{aligned}
\langle v(t)F(t') \rangle &= \int_{-\infty}^{\infty} \frac{d\omega}{2\pi} e^{-i\omega t} \int_{-\infty}^{\infty} \frac{d\omega'}{2\pi} e^{-i\omega' t'} \langle u(\omega)f(\omega') \rangle \\
&= \int_{-\infty}^{\infty} \frac{d\omega}{2\pi} e^{-i\omega t} \int_{-\infty}^{\infty} \frac{d\omega'}{2\pi} e^{-i\omega' t'} \frac{2\pi c_0 \delta\,(\omega + \omega')}{\gamma - im\omega} \\
&= ic_0 \int_{-\infty}^{\infty} \frac{d\omega}{2\pi} e^{-i\omega(t-t')} \frac{1}{m\omega + i\gamma} \\
&= \begin{cases} \dfrac{c_0}{m} e^{-\gamma(t-t')} & (t > t') \\[2mm] 0 & (t < t') \end{cases}
\end{aligned}
\tag{10.36}
$$

In the limit $t \to t'$, we take the average value

$$
\langle v(t)F(t) \rangle = \frac{c_0}{2m} \tag{10.37}
$$

The energy balance (10.34) now reads

$$
\frac{dK}{dt} = \frac{c_0}{2m} - \frac{2\gamma}{m} K \tag{10.38}
$$

with solution

$$
K = \frac{c_0}{4\gamma} \left(1 - e^{-2\gamma t}\right) \tag{10.39}
$$

The asypmtotic value $c_0/4\gamma$ reproduces the earlier result (10.11).

# Chapter 11

# The Life Process

## 11.1.　Life

According to the Chinese dictionary *Shuo Wen* （说文）, compiled by Xu Shen （许慎） in 101 AD, "life" connotes "advancement." Its hieroglyphic character symbolizes blades of grass pushing out of the earth, apparently driven by some incessant, irresistible force to maintain, promulgate and improve a highly complex organization. The stage on which the process plays out is the living cell, and the agents for cellular functions are the protein molecules.

進也
象艸木生出土上
説文解字第六篇下

## 11.2.   Cell Structure

Biological cells have sizes ranging from $10^{-7}$–$10^{-4}$ m, as shown in
Fig. 11.1 in comparison with other systems.

A cell in a living body is immersed in an aqueous environment,
with typical structure schematically shown in Fig. 11.2.

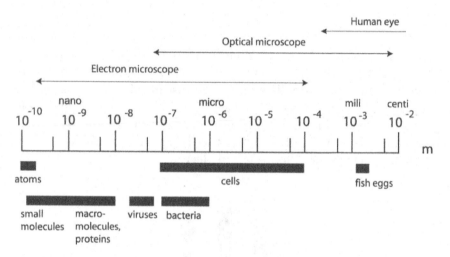

Fig. 11.1.   Log scale of comparative dimensions.

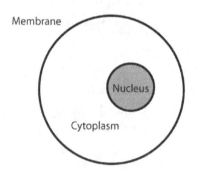

Fig. 11.2.   Schematic structure of a cell.

The stuff inside the cell membrane is called the **cytoplasm**, which is 95% water. The remainder consists of, by weight,

- 50% protein,
- 15% carbohydrate,
- 15% nucleic acid,
- 10% lipid,
- 10% misc.

The acidity of a solution is measured by

$$\text{pH} = -\log_{10}[\text{H}^+] \qquad (11.1)$$

where $[\text{H}^+]$ is the concentration of $\text{H}^+$ ions. For pure water $[\text{H}^+] = 10^{-7}$, giving pH = 7. Below this value a solution is consider an acid, and above this, a base. In a cell, the pH lies between 7.2–7.4.

Cells that have a nucleus are designated as *eukaryotic*, and this includes most plant and animal cells. The nucleus, bound by its own membrane wall, contains genetic molecules collectively known as *chromatin*. A primitive type of cell without a nucleus is called *prokaryotic*, an example of which is the single cell of the bacterium *E. Coli*. In this case, the chromatin floats free in the cytoplasm.

Among the chromatin are *chromosomes*, which in the human cell are very long and very thin threads, of diameter $4 \times 10^{-9}$ m and length 1.8 m. Each chromosome consists of one DNA molecule, festooned at regular intervals with bead-like proteins called *histones*. These help to wind the chromosome, through several hierarchies of twisting, into a tight knot that fits inside a nucleus of typical diameter $6 \times 10^{-6}$ m. In Fig. 11.3 a chromosome is shown at different stages of unwinding.

The DNA molecule can be divided into sequences called *genes*; one gene carries the code for making one type of protein. The *central dogma* of biology states that protein synthesis proceeds according to the sequence

$$\text{DNA} \underset{\text{transcription}}{\longrightarrow} \text{RNA} \underset{\text{translation}}{\longrightarrow} \text{Protein} \qquad (11.2)$$

The transcription occurs inside the nucleus, where a specific gene is copied to an *RNA* molecule, which is then transported through pores in the nuclear membrane to the cytoplasm. There they are translated into protein molecules at a *ribosome*, which is made up of a special RNA and 80 proteins. There are hundreds of thousands

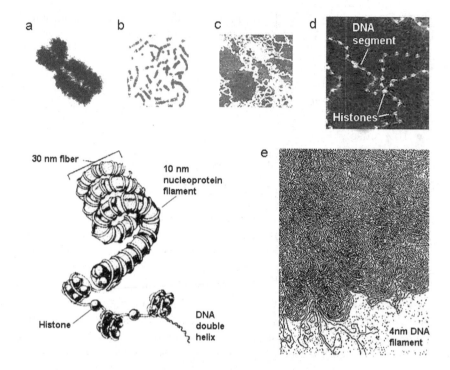

Fig. 11.3.    (a–e)    Chromosome in different stages of unwinding. Lower left panel is a schematic representation of the hierarchies of twisting, starting with the DNA double helix in the lower right corner. [Adapted from *MIT Biology Hyperbook*, http://www.botany.uwc.ac.za/mirrors/MIT-bio/bio/7001main.html]

of ribosomes floating freely in the cytoplasm, or attached to special molecular structures.

A great variety of proteins is needed to perform different functions in a living cell. Even in as simple a cell as *E. Coli* there are thousands of types of proteins.

## 11.3.  Molecular Interactions

There are different types of interaction between atoms and molecules. In biological studies, interaction energy is usually given in units of kcal/mole, with the following equivalences:

$$1\,\text{kcal/mole} = 0.0433\,\text{eV} = 503\,\text{K} \qquad (11.3)$$

For reference, thermal energy at room temperature $(T = 300\,\text{K})$ is

$$kT = 0.6\,\text{kcal/mole} \qquad (11.4)$$

- The *covalent bond* involves the sharing of two electrons between the interacting partners, with binding energy 50–150 kcal/mole. This is regarded as unbreakable at room temperature. An example is the water molecule $H_2O$, where the hydrogen atoms are bound to the oxygen via covalent bonds. The bonding leaves H with an excess positive charge, and O with an excess negative charge. This is indicated by the notation

$$H^{\delta+} - O^{\delta-} - H^{\delta+} \qquad (11.5)$$

- The *hydrogen bond* involves sharing an H atom between the interacting partners, with a binding energy of 5 kcal/mol. The bond has polarity, with H covalently bonded to one partner, and more weakly attached to the other, through its excess charge. Examples are

$$-O^{\delta-} - H^{\delta+} \cdots O^{\delta-} -$$
$$-O^{\delta-} - H^{\delta+} \cdots N^{\delta-} - \qquad (11.6)$$

where $\cdots$ denotes a hydrogen bond.
- The *ionic bond* arises from the exchange of one electron, with binding energy 4–7 kcal/mol. An example is the bond $Na^+$–$Cl^-$, which gives rise to the crystalline structure of salt.
- The *van der Waals interaction* arises from mutual electric polarization, and is relatively weak, amounting to 0.2–0.5 kcal/mol.

In addition, charged molecules interact through the electrostatic Coulomb potential screened by the surrounding solution.

## 11.4. Primary Protein Structure

A protein is a long-chain molecule consisting of a backbone made up of amino acids connected sequentially via a peptide bond. For this reason, the chain is called a *polypeptide chain*. The number of units (amino acids) on the chain ranges from the order of 50 to 3000. The amino acids are chosen from a list of 20, and the sequence of the amino acids is called the *primary structure*. A schematic diagram of the polypeptide chain and the peptide bond are given in the upper panels of Fig. 11.4. The amino acids are organized around a central carbon atom labeled $C_\alpha$, and differ from one another only in their side chains. Thus, we can picture the basic constitution of the protein molecule as different side chains attached to a "backbone." The repeating units along the main chain are referred to as "residues." This is illustrated in the lower panel of Fig. 11.4. The chemical composition of the 20 side chains, grouped according to their reaction to water (hydrophobic, hydrophilic, or in between) are shown in Fig. 11.5.

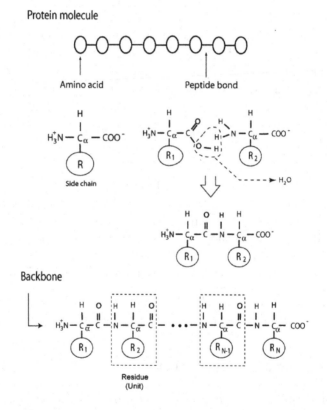

Fig. 11.4. Schematic representations of a protein molecule. [Partly adapted from *MIT Biology Hyperbook*, http://www.botany.uwc.ac.za/mirrors/MIT-bio/bio/7001main.html]

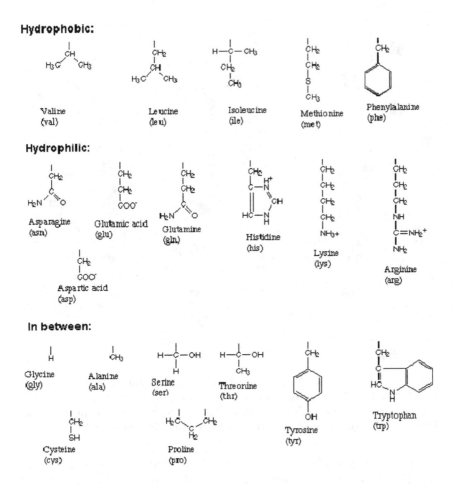

Fig. 11.5. The side chains of the 20 possible amino acids in a protein molecule. [Adapted from *MIT Biology Hyperbook*, http://www.botany.uwc.ac.za/mirrors/ MIT-bio/bio/7001main.html]

## 11.5. Secondary Protein Structure

Proteins are synthesized *in vivo* (in the cell) in molecular structures called *ribosomes*, and released into the cytoplasm. In this aqueous environment, it folds into its "native state," whose geometrical shape is important for biological functioning. Locally, the chain curls up into α-helices, or braids into β-sheets. These are called *secondary structures*.

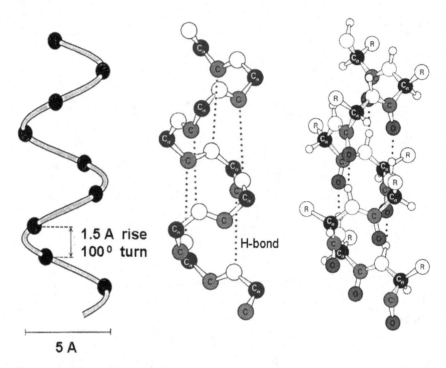

Fig. 11.6.  An α-helix in different schematic representations with increasing detail. The black spheres represent the central carbon atom in an amino acid. Dotted lines represent hydrogen bonds. [Adapted from *MIT Biology Hyperbook*, http://www.botany.uwc.ac.za/mirrors/MIT-bio/bio/7001main.html]

An alpha helix is shown in Fig. 11.6 in various views. It is held together by hydrogen bonds, with an $NH$ group as donor, and $C = O$ group as acceptor. Forming such a helix allows the hydrophobic side chains to be buried inside the folded molecule to avoid contact with water, and at the same time saturate the hydrogen bonds within the molecule, without reaching outside for a water molecule.

A β-sheet uses hydrogen bonding in a different manner, as illustrated in Fig. 11.7.

## 11.6.  Tertiary Protein Structure

The helices and sheets are arranged in a three-dimensional architecture called the *tertiary structure*. A striking difference between the

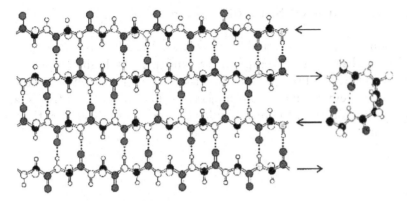

Fig. 11.7. A $\beta$-sheet, with one of the end groups is shown on the right. Dots represent hydrogen bonds. [From *MIT Biology Hyperbook*, http://www.botany.uwc.ac.za/mirrors/MIT-bio/bio/7001main.html]

secondary structures and the tertiary structure is the high degree of regularity in the former, and the irregular appearance of the latter. Kendrew, who discovered the tertiary structure of myoglobin through X-ray diffraction, was shocked at the irregularity, conditioned as he was by the earlier discovery of the DNA double helix to expect something of similar "beauty." His model of the molecule is shown in Fig. 11.8. However, the formation of secondary and tertiary structures are not independent of each other.

10 A

Fig. 11.8. Kendrew's model of myoglobin, in three different views. The sausage-shaped tubes are low-resolution representations of $\alpha$ helices. [From C. Branden and J. Tooze, *Introduction to Protein Structure*, 2nd ed. (Garland Publishing, NY, 1998), Fig. 2.1.]

## 11.7.   Denatured State of Protein

Proteins may be extracted from cells and studied *in vitro* (in test tube) under different temperatures and pH. At high temperatures and low pH, the protein unfolds into a random coil. Upon restoring the native environment, the molecule refolds into its native shape. Depending on the protein, the folding time ranges from less than 1 s to 15 min.

# Chapter 12

# Self-Assembly

## 12.1. Hydrophobic Effect

The dominant driving force for protein folding is the *hydrophobic effect*, which arises from the fact that water molecules seek to form hydrogens bonds with each other, or with other polar molecules. The presence of nonpolar molecules in water obstructs such bond formation, and the water molecules try to push them out of the way. The net effect is that the nonpolar molecules appear to avoid contact water — they are "hydrophobic." On the other hand, the polar molecules can form hydrogen bonds, are said to be "hydrophilic." As we have mentioned, protein folding is driven by the hydrophobic effect. A picture of a crambin protein molecule in water is shown in Fig. 12.1, which is reconstructed from X-ray diffraction data.[1] The water network "squeezes" the protein to maintain it in its native state.

Protein folding is an example of "self-assembly," in which components of the system organize themselves in response to features in the environment. Mutual interactions among the component play a role in the structure, but they are not the driving forces. An example is shown in Fig. 12.2.

Micelles and bilayers are examples of self-assembly driven by the hydrophobic effect, but are much simpler than protein molecules.

---

[1]M.M. Teeter, in *Protein Folding: Deciphering the Second Half of the Genetic Code*, eds. L.M. Gierasch and J. King (Am. Asso. Adv. Sci. Pub. No.89-18S, Washington, D.C., 1989) pp. 44–54.

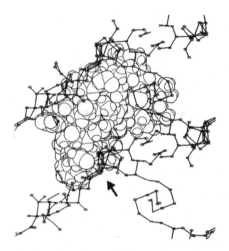

Fig. 12.1. Water molecules (small circles) adsorbed on the surface of a crambin molecule. Lines between water molecules indicate hydrogen bonds. Arrow indicates water accumulation at the position of an exposed hydrophobic residue.

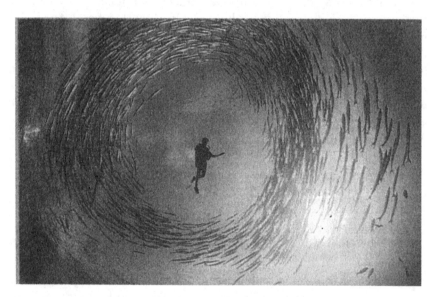

Fig. 12.2. Self-assemblage of baracudas around a diver. (Photograph taken in 1986 in Papua New Guinea by David Doubilet, photographer for *National Geographic Magazine*.)

## 12.2.   Micelles and Bilayers

Lipids, in particular phospholipids, are chain molecules with a hydrophilic "head," and hydrophobic hydrocarbon "tail," which may have one or more strands. Examples are shown in Fig. 12.3. Single-tail lipids are geometrically like cones, while those with more tails are like cylinders.

When lipid molecules are placed in water at sufficiently high concentrations, they aggregate to form *micelles* or *bilayers*, in order to hide the hydrophobic tails from water. The former is a sphere formed from cone-like lipids, with heads on the surface and tails inside. The latter is a two-layer structure made of cylinder-like lipids, with the heads forming the outer surfaces, and the tails sandwiched between the layers. These configurations are illustrated in Fig. 12.4. Bilayers exhibit a solid-like phase in which the lipids arrange themselves in a hexagonal lattice on the plane of the layer, and the tails of the

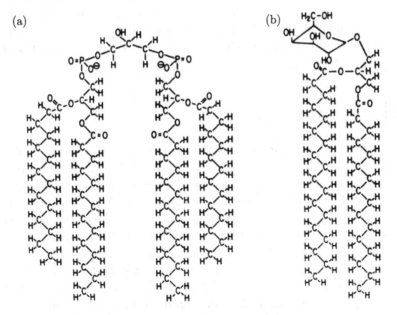

Fig. 12.3.   Phospholipids: (a) Cardiolipin, (b) Phophatidylinositol. [From P.L. Yeagle, *The Membranes of Cells*, 2nd ed. (Academic Press, NY, 1993).]

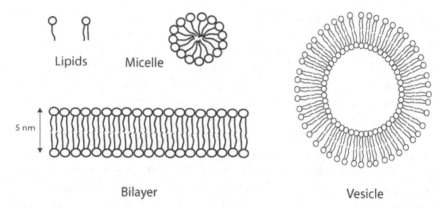

Fig. 12.4.   Micelles and bilayers formed by lipids in water. Right panel shows a vesicle — a sac with bilayer wall.

lipids are rigid. The thickness of the layer, typically 5 nm, is about twice the length of the lipid tails. At higher temperatures there is a first-order phase transition into a liquid-like phase with greater thickness, in which the lipids have considerable fluidity of motion in the plane of the layer. Micelles can engulf water-insoluble molecules, rendering the latter effectively water soluble. This is how soap works to get rid of oil stains.

## 12.3.   Cell Membrane

An actual cell membrane, called a *plasma membrane*, is a bilayer embedded with other molecules that serve specific purposes. For example, cholesterol molecules are embedded for mechanical strength. There are protein molecules as well that act as receptors for chemical signals, or create pores through which ions can pass. These adjuncts are integral parts of the plasma membrane.

A protein is embedded in the membrane by threading partially or completely through the bilayer, with its hydrophilic residues sticking outside the membrane, and the hydrophobic ones between the layers, as schematically illustrated in Fig. 12.5. The segment inside generally contains 20–30 residues. In this manner, a protein can fashion itself into a pore — a "barrel" with a molecular-sized

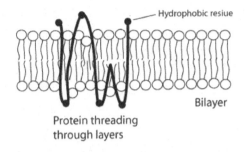

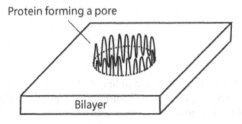

Fig. 12.5. (Upper) Protein molecule threading through a bilayer. (Lower) Protein molecule forming a barrel-shape pore, with molecular-size hole.

opening, as illustrated in Fig. 12.5. Selected ions are allowed to pass through these barrels, like ships passing through a dock of a canal.

Molecules can also pass in and out of the cell membrane via processes known as *endocytosis* (into cell) and *exocytosis* (out of cell). Endocytosis is depicted in the upper panel in Fig. 12.6, showing an amoeba cell swallowing a food particle. The membrane deforms to envelope the food particle, and the envelope then detaches from the membrane and becomes part of the cytoplasm. The sensing of the food particle, as well as the deformations of the membrane, are done with the help of embedded proteins.

Exocytosis is illustrated in the lower panel in Fig. 12.6. A structure in the cell called the *Golgi apparatus* synthesizes vesicles, complete with embedded proteins on its bilayer wall. These vesicles migrate to the cell membrane, and fuse with it, and the content of the vesicles (more proteins) are thereby secreted from the cell. This process also serves the important function of renewing the plasma membrane following endocytosis.

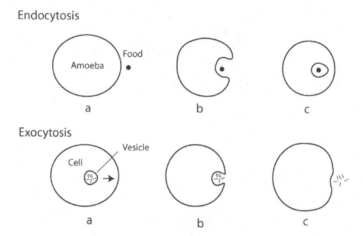

Fig. 12.6. Processes showing the dynamic nature of the cell membrane. (Upper) Endocytosis illustrated by amoeba engulfing a food particle. (Lower) Exocytosis illustrated by secretion of proteins from a cell.

## 12.4.  Kinetics of Self-Assembly

Micelles exhibit various phases, with interesting kinetics for their formation. The system consists of lipids in solution, and there are three parameters under the experimenter's control:

- temperature,
- lipid concentration,
- nature of the solution.

At a fixed temperature for low concentrations, the lipids form a "gas" of monomers. When the concentration is increased beyond a critical value called the *critical micellization concentration* (CMC), the lipid molecules assemble into micelles. The CMC can be varied by varying the nature of the solution. Above the CMC, there is another critical concentration at which the micelles form a crystal. The phase diagram resembles the familiar PT diagram of real gases, as shown in Fig. 12.7.

The nature of the solution also determines the shape of the micelle. In an aqueous mixture of lecithin and bile salt, for example, the micelle shape changes with dilution of the medium. At high strengths, the micelle is spherical. Upon dilution, the spheres

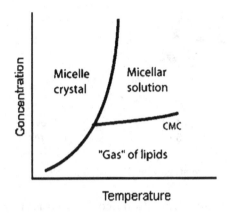

Fig. 12.7.   Phase diagram of micelles.

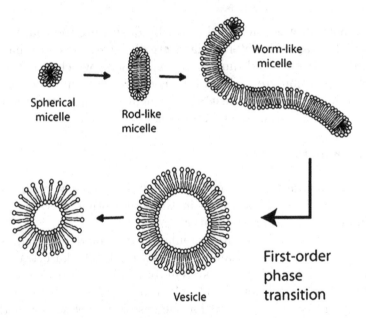

Fig. 12.8.   Micelle–vesicle transition upon dilution of the medium.

elongate into rods, then to worm-like shapes, and finally goes through a first-order phase transition to vesicles. Beyond this point, the vesicles contract upon further dilution. This is illustrated in Fig. 12.8.

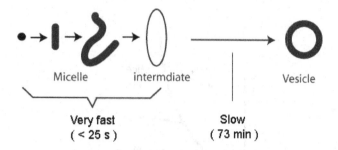

Fig. 12.9.   Two-stage kinetics of the micelle–vesicle transition.

The micelle–visicle transition occurs via a two-step process,[2] with a fast stage followed by a slow stage:

$$\text{Micelles} \xrightarrow[\text{very fast}]{} \text{Intermediates} \xrightarrow[\text{slow}]{} \text{Vesicles}$$

The transition was initiated by rapidly quenching the solution into the vesicle phase. The intermediates are disk-like structures that were formed within the dead time of the experiment, which was 25 s. The vesicle population grew with a time constant of 73 min. This is summarized in Fig. 12.9. The fast–slow sequence of phase transitions has also been observed in other experiments.[3,4]

## 12.5.   Kinetic Arrest

The two-stage fast–slow kinetics is a common feature of first-order phase transitions. This occurs in the long-lasting glass stage of a liquid on its way to crystallization.

It occurs in the computer simulation of spinodal decomposition discussed earlier. This example shows that local phase transitions occur rapidly throughout the system, but the coalescence into global phases takes a much longer time. It also suggests that the slowness means more time is required for the system to search for paths in phase space leading to the final state. The system may be said to be in "kinetic arrest."

[2]S.U. Engelhaaf and P. Schurtenberger (1999) *Phys. Rev. Lett.* **82**, 2804.
[3]H. Wang *et al.* (2003) *Phys. Rev.* **E67**, 060902(R).
[4]S.S. Funari and G. Rapp (1999) *Proc. Natl. Acad. Sci. USA* **96**, 7756.

This phenomenon was also observed in the crystallization of colloidal suspensions.[5] At a volume fraction 0.58 occupied by the colloidal particles, the system becomes a "glass" that takes months, even years to crystallize.[6] Yet, in microgravity, crystallization occurred in less than two weeks,[7] suggesting that the blockage in phase space can be removed by trivial changes.

---

[5]V.J. Anderson and H.N.W. Lekkerker (2002) *Nature* **416**, 811.

[6]P.N. Pusey and W. van Mergen (1986) *Nature* **320**, 340.

[7]J. Zhu *et al.* (1977) *Nature* **387**, 883. The experiment was performed in outer space on the space shuttle Columbia, where gravity was reduced to $10^{-6}$ g.

# Chapter 13

# Kinetics of Protein Folding

## 13.1. The Statistical View

In experiments on protein molecules, one studies samples containing a macroscopic number of molecules, either in solution or in crystalline form. The sample as a whole is of course a thermodynamic system with well defined properties, but a single protein molecule is too small to be considered a true thermodynamic system. Thermodynamic functions, such as the free energy, when indirectly deduced, are subject to large thermal fluctuations. Similarly, in the observation of protein folding, we can only measure the fraction of molecules having a certain average configuration, and there are large fluctuations about the average configuration. Thus, when we speak of the properties of a protein molecule, they must be understood in a statistical sense.

We should also bear in mind that properties we attribute to a single protein molecule, such as its free energy, depend strongly on the environment. It is the medium in which the molecule is immersed that induces the molecule to fold; it is the medium that maintains the native structure in a state of dynamic equilibrium. In addition to interactions among the residues, the free energy should reflect the hydrophobic interactions with the medium. It is expected to have many local minima, and the dynamic equilibrium state may be caught in one of these local minima, as determined by the kinetic path of the folding process. It may also switch rapidly among a group of neighboring minima.

There is a well-known "Levinthal paradox", which might be stated as follows: If a protein chain had to sample all possible configurations to reach the free-energy minimum, a simple calculation shows that the time required would be orders of magnitude greater than the age of the universe. How do actual protein molecules do it in a matter of minutes, even seconds?

The premises of the paradox are that (a) the folding process seeks a unique lowest minimum of the free energy; (b) any protein chain must fold into the state of minimum free energy. There is no reason to accept the assumptions. It is more likely that

- the free energy is not at an absolute, but at a constraint minimum, or a group of minima, as determined by the kinetics of folding;
- not all protein chains will fold, and of those that do, not all will fold in a short time. Only those protein chains that can fold in a short time were chosen by evolution to function in the living cell.

## 13.2. Denatured State

Protein molecules in a nonaqueous medium exists in an unfolded, or denatured state, which can be taken to be a random coil. Evidence for this is shown in Fig. 13.1.[1] The NMR[2] spectra of the protein lysozyme are compared to that of a random chain with the same primary structure. That is, the relative orientations between successive residues were chosen at random.

The polar residues in the protein chain seek to form hydrogen bonds, either with molecules in the medium, or with each other. Thus, $\alpha$-helices and $\beta$-sheets will form momentarily, but they are not permanent, since the bonds can rapidly switch from one molecule to another. The lifetimes of such bonds are probably of the order of $10^{-12}$ s, similar to those in the water network. These configurations are therefore random fluctuations.

---

[1]T.L. James, *Nuclear Magnetic Resonance in Biochemistry* (Academic Press, NY, 1997) p. 239.

[2]NMR (Nuclear Magnetic Resonance) measures the shift in the hyperfine structure of $H$ atom due to a change in its chemical environment. This "chemical shift" is given in terms of the frequency shift, or in ppm (parts per million).

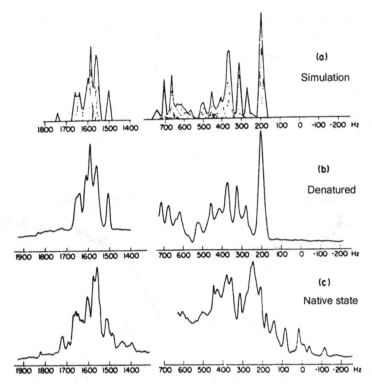

Fig. 13.1. NMR spectrum of hen lysozyme in $D_2O$ solution. (a) Simulation of random coil; (b) denatured state; (c) native state. Vertical axis is intensity in arbitrary units, horizontal axis is frequency shift in Hz.

## 13.3. Molten Globule

When the pH or temperature is changed, so that the medium becomes effectively aqueous, the protein chain begins to fold. In Fig. 13.2, the folding of lysozyme was followed in real time for about 1 s, employing techniques that measure various properties of the molecule.

- The permanent hydrogen bonds in $\alpha$-helices and $\beta$-sheets grow with time constants of approximately 50 and 200 ms, respectively.
- The helical content, a measure of secondary structure, first overshoots, and then decays to the native value.

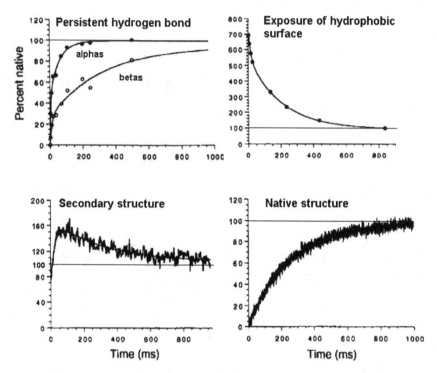

Fig. 13.2. Development of various features of hen lysozyme, during the folding process. [Adapted from C.M. Dobson, A.E. Phillip and E.R. Sheena (1999) *Trends Biochem. Sci.* **19**, 31.]

- The decrease in the exposure of hydrophobic surface has roughly the same time constant as for the decay of helical content. This indicates that fluctuations in the secondary structure become stabilized when hydrophobic residues become buried.

The folding time for the protein DHFR is much longer. Figure 13.3 displays the NMR spectrum as a function of time. We can follow the decay of the initial random coil, and the growth of native structures. Native features become recognizable after 20 s, but the completion of the folding process takes the order of 10 minutes.

These measurements do not reveal what happens at very short times. To find out whether there are stable structures near the native state, we turn to specific heat measurements. The specific heat capacity of canine milk lysozyme are shown as functions of

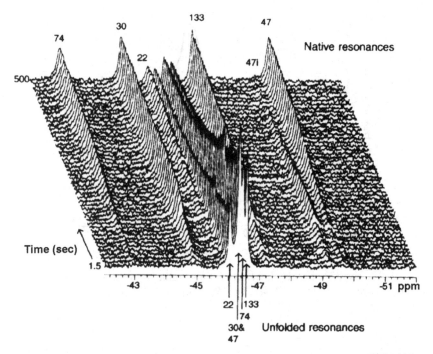

Fig. 13.3. Time evolution of the NMR spectrum of the protein DHFR (dihydrofolate reductase). [From S.D. Hoeltzli and C. Frieden (1998) *Biochemistry* **37**, 387.]

temperature in Fig. 13.4. What is actually plotted is the excess heat capacity of the solution when the proteins are added. The peak at a higher temperature marks a transition from the denatured state to an intermediate state known as the *molten globule,* and the second peak marks the transition to the native state. If the protein molecule had been infinitely large, we might imagine that these peaks would become delta functions, signifying first-order phase transitions. As they stand, we may associate them with pseudo phase transitions, and estimate the latent heats from the areas under the peaks. The results are given in the table below:

| | Denatured → Molten globule | Molten globule → Native |
|---|---|---|
| Latent heat | 100 kcal/mole | 120 kcal/mole |

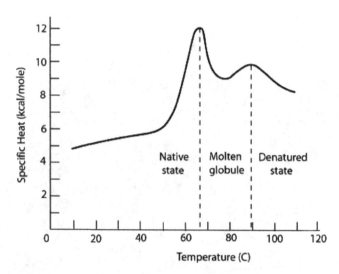

Fig. 13.4. Excess specific heat of canine milk lysozyme. [Redrawn from T. Koshiba *et al.* (1999) *Protein Eng.* **12**, 429.]

These experiments indicate that protein folding goes through two stages: a fast and a slow one. The fast stage lasts the order of $10^{-3}$ s, during which the denatured state becomes a molten globule. In the next stage, the molten globule slowly evolves into the native state:

$$\text{Denatured state} \xrightarrow{\text{fast}} \text{Molten globule} \xrightarrow{\text{slow}} \text{Native state}$$

The durations of the slow stage can be estimated from Fig. 13.2 for lysozome, and Fig. 13.3 for DHFR, and are summarized in the following table:

| Protein | Residue number | Fold time (s) |
|---------|---------------|---------------|
| Lysozyme | 164 | 1 |
| DHFR | 384 | 250 |

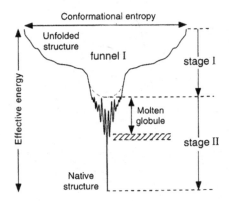

Fig. 13.5. The folding funnel. Horizontal dimension represents entropy, and vertical dimension represents energy.

## 13.4. Folding Funnel

The two stages in folding may be indicated symbolically in a "folding funnel", as shown in Fig. 13.5. During the first stage, the protein rapidly shrinks from a random coil to a molten globule, losing entropy with relatively small change in the energy. In the second stage, the molten globule slowly transforms into the native state, with little change in entropy, but large loss of energy.

The profile of the folding funnel is a plot of the enthalpy $H(P, S)$ as function of $S$ at constant $P$. The enthalpy is given by the thermodynamic formula

$$H = \int C_P \, dT \tag{13.1}$$

We can perform the integration graphically from Fig. 13.4, and then rescale the $T$ axis to convert it to $S$. The result is shown in Fig. 13.6, where the rescaling from $T$ to $S$ could be done only qualitatively.

## 13.5. Convergent Evolution

It is a remarkable empirical fact that many protein molecules share similar native folds. More specifically, proteins with sequence similarity below the level of statistical significance (as low as 8–9%) have

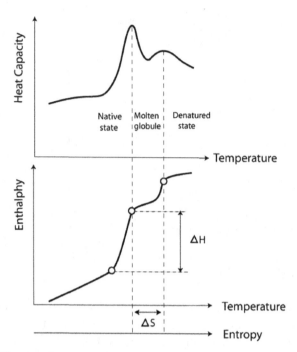

Fig. 13.6.   Plotting the enthalpy as a function of $S$ at fixed $P$ gives a quantitative profile of the folding funnel.

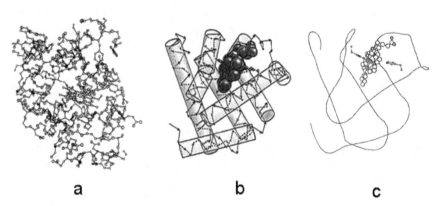

Fig. 13.7.   The globin fold: (a) with all atoms displayed; (b) with side-chains omitted, and the eight $\alpha$-helices shown as tubes; (c) skeleton of the tertiary structure. [Adapted from C. Branden and J. Tooze, *Introduction to Protein Structure*, 2nd ed. (Garland Publishing, NY, 1991), Fig. 2.9.]

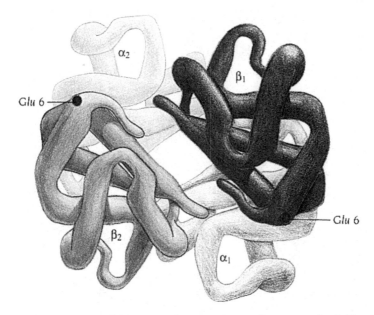

Fig. 13.8. The hemoglobin protein contains subgroups of globin folds. (C. Branden and J. Tooze, *op. cit.*, Fig. 3.13.)

similar three-dimensional structures and functions.[3] This property is known as "convergent evolution." This must mean that the molten globule state already has a degree of universality.

An example is the globin fold, as illustrated in Fig. 13.7, which is found in a large group of proteins, including myoglobin illustrated in Fig. 12.8. The larger protein hemoglobin is made up of four globin folds, as shown in Fig. 13.8.

---

[3]N.V. Dokholyan, B. Shakhnovich and E.I. Shakhnovich (2002) *Proc. Natl. Acad. Sci. USA* **99**, 14132.

# Chapter 14

# Power Laws in Protein Folding

## 14.1.  The Universal Range

For length scales in which the correlation function exhibits power-law behavior, the properties of a protein molecule should exhibit universal behavior, for the correlation length is effectively infinite, and details of the primary structure are not relevant. We call this the "universal range."

Consider the correlation function $g(r)$ of a protein, which gives the probability of finding a residue at $r$, when it is known that one is at the origin. We envision the limit in which the number of residues and the average radius tend to infinity. Let $\eta$ be the scale of atomic distances in the molecule, and $\xi$ the correlation length.

- On small length scales, $g(r)$ has fluctuations over distances of order $\eta$. We shall smooth out such fluctuations through spatial averaging. In experiments, the averaging is done through the finite resolution of the measuring instruments.
- On large length scales, $g(r) \to 0$ for $r \gg \xi$.
- The *universal range* lies in between, where $g(r)$ obeys a power law:

$$\eta \ll r \ll \xi$$
$$\eta^{-1} \gg k \gg \xi^{-1} \tag{14.1}$$

Since $r \ll \xi$, we are in the critical region where detailed structures of the system become irrelevant.

These properties are illustrated in sketches of $g(r)$ and its Fourier transform $\tilde{g}(k)$ in Fig. 14.1.

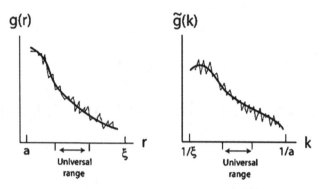

Fig. 14.1. Universal range, where the correlation function has power-law behavior.

## 14.2. Collapse and Annealing

The native protein state is as compact as ordinary matter, with a characteristic density. This corresponds to a compactness index $\nu = 1/3$, as shown in the plot of empirical data[1] in Fig. 14.2. Thus, in the universal range, the native state exhibits the power law

$$\tilde{g}(k) \sim k^{-3} \quad \text{(Native state)} \tag{14.2}$$

which has been directly verified in scattering experiments, as shown in Fig. 14.3.[2]

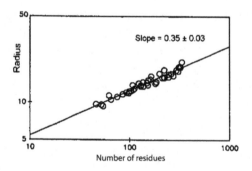

Fig. 14.2. Log–log plot of protein radius $R$ versus number of residues $N$, for 43 proteins, showing $R \sim N^{1/3}$.

[1]T.G. Dewey (1993) *J. Chem. Phys.* **98**, 2250.

[2]The experiment was reported in G. Damaschun *et al.* (1986) *Int. J. Biol. Macromol.* **13**, 226. The data was taken from O. B. Pittsyn, The molten globule state, in *Protein Folding*, ed. T. E. Creighton (Freeman, NY, 1992).

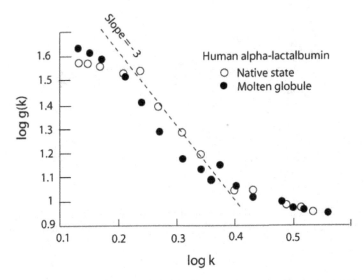

Fig. 14.3. Fourier transform of correlation function for a range of high wave numbers, in the native state and the molten globule state of a protein. The region of the −3 power law corresponds to the "universal range" described in the text.

Data for the molten globule are included in the same plot, and they show the same behavior:

$$\tilde{g}(k) \sim k^{-3} \quad \text{(Molten globule)} \qquad (14.3)$$

This shows that the molten globule is as compact as the native state.

In the denatured state, on the other hand, the protein is a random coil. In the universal range, where local structures are irrelevant, the structure is similar to a random polymer chain, for which the compactness index is well approximated by $\nu = 3/5$, and thus

$$\tilde{g}(k) \sim k^{-5/3} \quad \text{(Denatured state)} \qquad (14.4)$$

This result can be derived in a model based on the self-avoiding walk (SAW), which we shall describe later, and is verified by X-ray scattering from polymers.[3] It has the same exponent as Kolmogorov's famous 5/3 law in turbulence, which places the denatured protein

---

[3]K. Okano, E. Wada and H. Hiramatsu (1974) *Rep. Prog. Polym. Sci. Japan* **17**, 145.

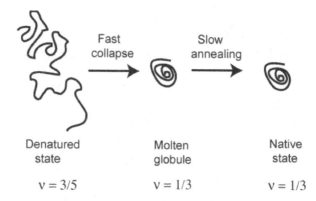

Fig. 14.4.    Fast collapse and slow annealing in protein folding. The compactness index $\nu$ changes from 3/5 to 1/3, the latter being the same as that for ordinary matter.

and turbulence in the same universality class. We shall expand on this relation later.

In summary, the folding process consists of a fast collapse, followed by slow annealing, as illustrated in Fig. 14.4.

## 14.3.    Self-Avoiding Walk (SAW)

Flory[4] models a random polymer chain by self-avoiding walk (SAW), which is random walk that is prohibited from revisiting an old position. It simulates the condition that two different monomers cannot occupy the same space. Other interactions among monomers are ignored.

It is easy to generate a sequence of SAW steps on a computer. A straightforward algorithm is to first generate a random step, and accept it if an old site is not revisited, and generate another step otherwise. One repeats the process until $N$ acceptable steps are obtained, and this represents one state of a polymer chain. By generating a sufficient number of chains independently, one has an ensemble of polymer chains, in which statistical averages can be calculated. It is harder to simulate the dynamics — the Brownian motion of the

---

[4]P. Flory, *Principles of Polymer Chemistry* (Cornell University Press, London, 1953).

chain. The most successful algorithm appears to be the "pivoting method," in which one rotates part of a chain about an arbitrarily chosen monomer, and accept the outcome only if it is self-avoiding.[5] A SAW of 1000 steps on a cubic lattice, is shown in Fig. 14.5 in stereoscopic views.

The compactness index can be calculated approximately, using an intuitive argument due to Flory. Consider a chain of $N$ monomers, with end-to-end distance $R$ in $D$-dimensional space, in the limit $N \to \infty$, $R \to \infty$. The monomer density is given by

$$n \sim \frac{N}{R^D} \tag{14.5}$$

The energy of the chain should contain a repulsive term due to self-avoidance, with energy density $\varepsilon$ proportional to $n^2$, the number of pairs per unit volume:

$$\varepsilon \sim kT v_0 n^2 \tag{14.6}$$

where $v_0$ is the excluded volume, and $kT$ is the thermal energy. The total repulsive energy is

$$E_{\text{repulsive}} \sim \varepsilon R^D \sim \frac{kT v_0 N^2}{R^D} \tag{14.7}$$

The repulsive energy tends to expand the volume at small $R$. As $R$ increases, however, the collision between two monomers becomes increasingly rare, and the sequence of steps should approach a simple random walk. This effect is taken into account by postulating a linear restoring force, corresponding to an "elastic energy" proportional to $R^2$:

$$E_{\text{elastic}} \sim k_B T \left( \frac{R}{R_0} \right)^2 \sim \frac{kT R^2}{N a^2} \tag{14.8}$$

where $R_0 = a\sqrt{N}$ is the characteristic length in simple random walk. The total energy is of the form

$$E = E_{\text{repulsive}} + E_{\text{elastic}} \sim kT \left( \frac{v_0 N^2}{R^D} + \frac{R^2}{N a^2} \right) \tag{14.9}$$

---

[5]B. Li, N. Madras and A.D. Sokal (1995) *J. Stat. Phys.* **80**, 661.

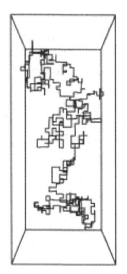 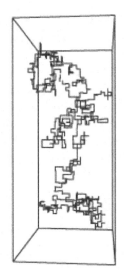

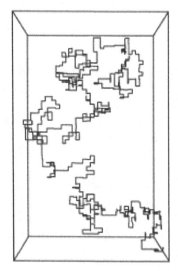 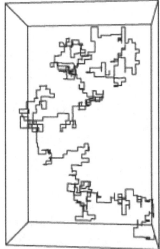

Fig. 14.5.   Stereoscopic views of SAW on a cubic lattice. Upper and lower panels represent front and side views. (Computer data courtesy Prof. Thomas Kennedy, Department of Mathematics, University of Arizona.)

Minimizing this with respect to $R$, we obtain the equilibrium radius

$$R_{\text{eq}} \sim N^{3/(D+2)} \tag{14.10}$$

The compactness index is therefore

$$\nu = \frac{3}{D+2} \tag{14.11}$$

For $D = 3$ this gives $\nu = 3/5 = 0.6$, which is remarkably close to the best value from computer simulations[6]

$$\nu = 0.5877 \pm 0.0006 \tag{14.12}$$

---

[6]B. Li, N. Madras and A.D. Sokal, *op. cit.*

# Chapter 15

# Self-Avoiding Walk and Turbulence

## 15.1. Kolmogorov's Law

The turbulent flow of a fluid is a stochastic process with dissipation. In the steady state known as fully developed homogeneous turbulence, the energy spectrum $E(k)$ obeys Kolmogorov's law in a range as of wave numbers $k$ called the "inertial range":

$$E(k) \sim \epsilon^{2/3} k^{-5/3} \qquad (15.1)$$

where $\epsilon$ is the rate of energy dissipation. The inertial range is what we call the universal range:

$$\eta^{-1} \gg k \gg \xi^{-1} \qquad (15.2)$$

where $\eta$ is a small length scale at which dissipation takes place, and $\xi$ is a correlation length.

The Kolmogorov law is based on general properties of the Navier–Stokes equation of hydrodynamics, physical assumptions and dimensional analysis.[1] It is universal, in the sense that it is independent of initial conditions, and independent of the detailed nature of the system. Experimental verification in different systems are shown in Figs. 15.1 and 15.2.

## 15.2. Vortex Model

A turbulent fluid can be modeled by a tangle of vortex lines. A vortex either terminates on an external wall or ends on itself to form

---

[1] C.C. Lin and W.H. Reid, Turbulent flow, theoretical aspects in *Hanbuch der Physik, Fluid Dynamics {II}*, eds. S. Flugge and C. Truesdell (Springer, Berlin, 1963). Reproduced in *Selected Papers of C.C. Lin*, Vol. 1, *Fluid Mechanics*, eds. D. Benny, F.H. Shu and C. Yuan (World Scientific, Singapore, 1987) pp. 175–260.

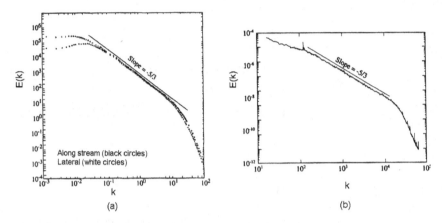

(a)                                    (b)

Fig. 15.1.  Energy spectra (in arbitrary units): (a) in jets. [Adapted from F.H. Champagne (1978) *J. Fliud Mech.* **86**, 67.], (b) in low-temperature heluim gas flow. [Adapted from J. Maurer, P. Tabeling and Zocchi. (1994) *Europhys.* **26**, 31.]

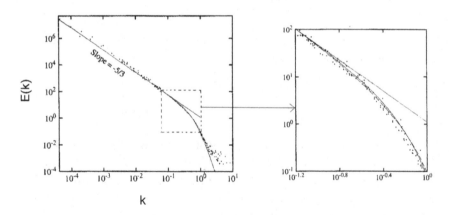

Fig. 15.2.  Energy spectrum (in arbitrary units), in tidal channel flow, showing details of transition from universal to dissipation range. [Adapted from H.L. Grant, R.W. Stewart and A. Millet (1962) *J. Fluid Mech.* **12**, 241.]

a ring. Kelvin's theorem in hydrodynamics states that, in the absence of viscosity, vortex lines do not cross. Thus the motion of a vortex line can be modeled by SAW. We are interested in situations with a high local density of lines, and it matters little whether we have a single long vortex line or a dense collection of separate lines. This immediately establishes an equivalence between turbulence and the polymer chain, and the exponent $-5/3$ in (14.1). With specialization, the model also accounts for the exponent $-2/3$.

Our model must break down at some small distance $\eta$ from the center, which defines a core radius. We shall regard the vortex line as a tube of radius $\eta$, with an intrinsic circulation

$$\kappa_0 = \oint_C \mathbf{v} \cdot d\mathbf{s} \tag{15.3}$$

where $\mathbf{v}$ is the velocity field, and $C$ is a circle of radius $\eta$ about the vortex line. We assume that energy dissipation occurs only at the vortex core. On length scales much larger than $\eta$, the vortex lines become indistinct. The number of vortex lines threading through a closed loop will be taken as the circulation around that loop divided by $\kappa_0$.

The correlation function $g(r)$ is now defined as the vortex line density at position $r$. According to classical hydrodynamics, a vortex line has an energy per unit length $\rho\kappa_0^2$, where $\rho$ is a mass density. Thus, the energy density at $r$ is given by

$$W(r) = \rho\kappa_0^2 g(r) \tag{15.4}$$

The Fourier transform gives the energy spectrum:

$$E(k) = \rho\kappa_0^2 \tilde{g}(k) \tag{15.5}$$

From our discussion of SAW earlier, $\tilde{g}(k) \sim k^{-5/3}$ in the universal range. Therefore

$$E(k) \sim \rho\kappa_0^2 k^{-5/3} \tag{15.6}$$

This is purely an expression of the correlation properties of vortex lines.

We can express $\kappa_0$ in terms of physical properties of the system. The rate of energy dissipation in hydrodynamics is given by

$$\epsilon \sim \nu\kappa_0^2 \tag{15.7}$$

where $\nu$ is the viscosity. We can eliminate $\nu$ in terms of the Reynolds number $R_0 = \lambda u/\nu$, where $\lambda$ is a length scale, and $u$ a typical velocity at that scale. With the assumption that energy dissipation occurs only at the vortex core, we take $\lambda u = \kappa_0$, and obtain

$$\epsilon \sim \frac{\kappa_0^3}{R_0} \tag{15.8}$$

or

$$\kappa_0 \sim (R_0 \epsilon)^{1/3} \tag{15.9}$$

Substituting this result into (15.6), and dropping constant factors, we obtain Kolmogorov's law

$$E(k) \sim \epsilon^{2/3} k^{-5/3} \tag{15.10}$$

The exponent 5/3 is universal, because it is a property of SAW. On the other hand, the exponent 2/3 is not universal, for it depends on physical properties of hydrodynamic vortices.

## 15.3.  Quantum Turbulence

The vortex model also applies to superfluid turbulence in a quantum fluid. The difference from the classical case is that vorticity is quantized, and we must take

$$\kappa_0 = \frac{h}{m} \tag{15.11}$$

where $h$ is Planck's constant, and $m$ is the mass of a particle in the fluid.[2] Kelvin's theorem no longer holds, and vortex lines do cross. When that happens, however, they immediately reconnect to form new vortex lines.[3] Thus, effectively, the lines do not cross, and we can use the SAW model. This places superfluid turbulence in the same universality class as classical turbulence and the polymer chain. The 5/3 law in superfluid turbulence has been found to be consistent with experiments.[4]

---

[2]R.J. Donnelly, *Quantized Vortices in Helium II* (Cambridge University Press, Cambridge, 1991).

[3]J. Koplik and H. Levine (1996) *Phys. Rev. Lett.* **76**, 4745.

[4]S.R. Stalp, L. Skrbek and R.J. Donnelly (1999) *Phys. Rev. Lett.* **82**, 4831.

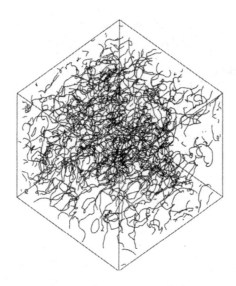

Fig. 15.3. Computer simulation of a vortex tangle in quantum turbulence. It is a fractal of dimension 1.6.

Figure 15.3 shows computer simulation of a vortex tangle in a quantum fluid,[5] which is a fractal of dimension 1.6.

## 15.4. Convergent Evolution in Turbulence

The energy spectrum in turbulence quickly reaches the universal Kolmogorov form, independent of the type of fluid or initial conditions. This is an example of convergent evolution. The general picture that emerges from long periods of theoretical and semi-empirical analysis may be summarized as follows:

- Energy is supplied to the flow via large-scale structures, such as a grid in a wind tunnel, boundary layers, wakes, weather fronts and the jet stream.
- The energy "cascades" down through successively smaller length scales, with negligible dissipation. The mechanism for the cascade is the breakup of large eddies into successively smaller ones.

---

[5]D. Kivotides, C.F. Barenghi and D.C. Samuels (2001) *Phys. Rev. Lett.* **87**, 155301.

- The energy is dissipated at a minimal length scale, where it is dissipated via viscosity, as soon as it is received.

The energy spectrum can be modified by a change in global conditions. For example, the turbulence in a rotating fluid has an energy spectrum with exponent between $-5/3$ and $-2$, depending on the angular velocity.[6]

---

[6]Y. Zhou (1995) *Phys. Fluids* **7**, 2092.

# Chapter 16

# Convergent Evolution in Protein Folding

## 16.1. Mechanism of Convergent Evolution

Convergent evolution refers to the property that a dynamical system reaches a steady state independent of initial conditions. C.C. Lin has emphasized that, on empirical evidence, this is a general phenomenon in dissipative stochastic systems.[1] Can we understand the physical basis for this?

To maintain equilibrium at a given temperature, a system must exchange energy with its environment. Consider, for example, an atom in a gas. Treating the rest of the gas as the environment, we can describe its dynamics through the Langevin equation. The energy input comes from random forces, and dissipation occurs due to friction. In order to reach thermodynamic equilibrium, both forces must be present, and related by the fluctuation–dissipation theorem. The maintenance of the equilibrium Maxwell–Boltzmann velocity distribution requires energy absorption and dissipation by the atom.

The example of a single atom is simple, in that the system has no intrinsic length scale. In systems with a large number of length scales, such as a turbulent fluid, or a long-chain molecule, the mechanisms of energy input and dissipation are more complex. We suggest, in analogy with turbulence, that the general mechanism for energy exchange involves the energy cascade:

- Energy is absorbed by the system at some input length scale.
- The energy is transferred to successively shorter length scales. This happens because the number of degrees of freedom generally increases as the length scale is decreased. Spreading an amount of energy among different degrees of freedom increases the entropy.

---

[1]C.C. Lin (2003) On the evolution of applied mathematics, *Acta Mech. Sin.* **19**(2), 97–102.

- The cascade ends at a microscopic scale, where the energy is dissipated as heat, through atomic processes. Depending on the system, dissipation can occur on all scales at varying degrees, or, as in the case of turbulence, only at the smallest scale.

The manner in which energy is being transferred from one degree of freedom to another may differ in different systems, but it is independent of initial conditions. Thus, after a few steps in the cascade, we expect all memory of the initial state to be lost.

In the absence of a complete theory, however, the scenario suggested above must be taken as a conjecture. An example of this as a simple model is given in Appendix A.

## 16.2. Energy Cascade in Turbulence

Let us review the mechanism for energy cascade in turbulence, which is well-understood:

- Energy input occurs at large length scales, through the creation of large vortex rings.
- The vortex rings are unstable because of "vortex stretching," i.e. the core of the vortex spontaneously contracts to a smaller radius, and increases in length. Eventually the vortex ring breaks up into smaller rings.
- The vortex ring ceases to exist as such, when its radius becomes comparable to the core radius. Its energy is then dissipated as heat.

The steady state in turbulence consists of a distribution of vortex rings of all sizes, from the input scale to the dissipation scale. The distribution is independent of initial conditions, for it depends only on the nature of vortex instability. A schematic representation of the energy cascade in turbulence is shown in Fig. 16.1.

## 16.3. Energy Cascade in the Polymer Chain

Since both turbulence and the random polymer chain (which is a model for the denatured protein state) can be modeled by SAW, we can envision an energy cascade in the latter. The energy in this case is

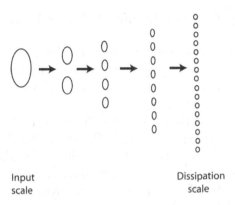

Input                              Dissipation
scale                                 scale

Fig. 16.1.   Energy cascade in turbulence. Large vortex rings break up into smaller ones, until dissipated at a small scale. In steady state, there is a distribution of vortex rings of all sizes.

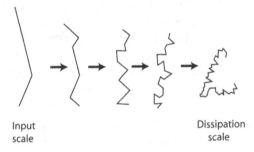

Input                              Dissipation
scale                                 scale

Fig. 16.2.   Energy cascade in a segment of a random polymer chain.

the kinetic energy of motion. The process is represented symbolically in Fig. 16.2, with the main points summarized as follows:

- Through concerted action of atoms in the medium, a portion of the chain gets bent on a large length scale. Dissipation is negligible, because the velocities involved are small. On the other hand, significant kinetic energy is imparted to the system through the induced mass motion.

- The energy cascades down the length scales, through coupling between the normal modes of the chain.

- The energy is dissipated at a microscopic scale, where velocities become large, and viscosity becomes important.

## 16.4. Energy Cascade in the Molten Globule

In an aqueous medium, the protein chain begins to fold, driven by hydrophobic forces. It rapidly collapses into a molten globule, with release of latent heat. In this process, both the secondary and tertiary structures must be considered.

The latent heat arises from the formation of semi-permanent bonds in the secondary structures. However, this bonding is insufficient to bind the molten globule, as shown by the fact that the latter cannot be maintained in a nonaqueous medium. Its stabilization depends on the effective pressure from hydrophobic forces, which arises from the desire of hydrophobic residues to be buried, and thus shielded from contact with water. Water nets would form around the molten globule to confine it.

The latent heat is released to the environment through the surface of the collapsing structure. The molten globule is as compact as the native state, and the surface-to-volume ratio is greatly reduced. That is why it will take a much longer time to release the latent heat, in the transition to the native state.

We suggest the following scenario for the energy cascade that maintains the molten globule as a semi-equilibrium state:

- Energy input occurs on the largest length scale, i.e. the overall dimension of the molecule. The water nets have a range of oscillation frequencies. The one that resonates with the lowest frequency of the structure transfers energy to that mode, in a manner similar to that in a driven harmonic oscillator in Brownian motion. These frequencies are of the order of 10 GHz.
- The energy is either
  — transferred to smaller length scales with no dissipation, through nonlinear structural interactions, or
  — converted into heat through internal friction: The friction arises from the unfolding–refolding of helices in the interior due to thermal fluctuations. This dissipates energy, because statistically a helix cannot refold exactly back to its original state.
- When the energy reaches the smallest scales corresponding to the radii of helices and loops exposed on the surface, it is dissipated into the medium.

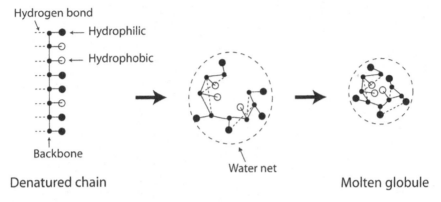

Fig. 16.3. Energy cascade in the molten globule state. Energy input occurs at large length scale, from the hydrophobic driving force. Dissipation occurs at small scales, at the surface of the protein.

The scenario is symbolically represented in Fig. 16.3.

In the meantime, bonding in the secondary elements continues, accompanied by release of latent heat, and the molten globule slowly transforms into the native state.

Energy exchange with the environment continues in the native state, and the energy cascade continues; but the rate of energy transfer is much reduced, because there is no latent heat to be dissipated.

## 16.5. Secondary and Tertiary Structures

What principle guides the formation of the molten globule? It is important to understand this, for the molten globule already possesses the basic structure of the native state.

We can only offer to guess this principle, namely that it is a combination of

- the minimization of local free energy in the formation of secondary structures, and
- the minimization of the time for energy cascade in the formation of the tertiary structure.

These two mechanisms influence each other, and are inseparable. We still lack any mathematical model, even a simple one, to describe the interplay.

# Appendix A

# Model of Energy Cascade in a Protein Molecule

## A.1.  Brownian Motion of a Forced Harmonic Oscillator

A harmonic oscillator is a model for a normal mode of a system, such as a protein molecule. Imagine that the molecule is immersed in water. It experiences dissipative and random forces. In addition, it is encaged in a water net, which exerts an organized oscillatory force on the molecule, and can transfer energy to it. Thus, the forced oscillation should alter the fluctuation–dissipation theorem.

Consider the Langevin equation of a harmonic oscillator of mass $m$ and natural frequency $\omega_0$:

$$m\ddot{x} + m\omega_0^2 x + \gamma\dot{x} = G(t) + F(t) \tag{A.1}$$

where $G(t)$ is a sinusoidal driving force with Fourier transform $g(\omega)$:

$$\begin{aligned} G(t) &= b\cos\omega_1 t \\ g(\omega) &= \pi b[\delta(\omega - \omega_1) + \delta(\omega + \omega_1)] \end{aligned} \tag{A.2}$$

and $F(t)$ is the random force with Fourier transform $f(\omega)$ satisfying

$$\begin{aligned} \langle f(\omega) \rangle &= 0 \\ \langle f(\omega)f(\omega') \rangle &= 2\pi c\delta(\omega + \omega') \\ \langle f(\omega)g(\omega') \rangle &= 0 \end{aligned} \tag{A.3}$$

where $\langle\rangle$ denotes ensemble average. Multiplying both sides of (A.1) by $\dot{x}$, we have

$$m\dot{x}\ddot{x} + m\omega_0^2\dot{x}x + \gamma\dot{x}^2 = \dot{x}(G + F)$$

$$\frac{d}{dt}\left(\frac{m}{2}v^2 + \frac{m\omega_0^2}{2}x^2\right) = v(G + F) - \gamma v^2 \tag{A.4}$$

Taking ensemble averages, we obtain the equation for energy balance:

$$\frac{dK}{dt} + \frac{dP}{dt} = \langle vF\rangle + \langle vG\rangle - 2\kappa K$$

$$K = \frac{m}{2}\langle v^2\rangle \tag{A.5}$$

$$P = \frac{m\omega_0^2}{2}\langle x^2\rangle$$

with

$$\kappa = \frac{\gamma}{m} \tag{A.6}$$

We can make the following identifications:

- Rate of work done on system $= \langle vF\rangle + vG$
- Rate of energy dissipation $= 2\kappa K$

The Fourier transform of $x(t)$ is readily found to be

$$z(\omega) = -\frac{g(\omega) + f(\omega)}{m(\omega^2 - \omega_0^2 + i\omega\kappa)} \tag{A.7}$$

We consider the modification of the fluctuation–dissipation theorem compared to the free-particle case, and will quote results without derivation. The following relations are relevant to energy balance:

$$\frac{d\bar{P}}{dt} = 0$$

$$\langle v(t)F(t)\rangle = \frac{c}{2m} \tag{A.8}$$

$$\overline{\langle v(t)G(t)\rangle} = \frac{\kappa b^2\omega_1^2}{2}\frac{1}{\left(\omega_0^2 - \omega_1^2\right)^2 + \omega_1^2\kappa^2}$$

where an overhead bar denotes averaging over a time long enough to be compared to the period of the driving force. Thus

$$\frac{d\bar{K}}{dt} = \frac{c}{2m} + \frac{\kappa b^2 \omega_1^2}{2m} \frac{1}{\left(\omega_0^2 - \omega_1^2\right)^2 + \omega_1^2 \kappa^2} - 2\kappa\bar{K} \qquad (A.9)$$

After a sufficiently long time, when steady-state motion is established, we should have $d\bar{K}/dt = 0$, and

$$\bar{K} = \frac{T}{2} \qquad (A.10)$$

where $T$ is the absolute temperature in energy units.[1] We then obtain the fluctuation–dissipation theorem

$$\frac{c}{2m\kappa} + \frac{b^2\omega_1^2}{2m} \frac{1}{\left(\omega_0^2 - \omega_1^2\right)^2 + \omega_1^2\kappa^2} = T \qquad (A.11)$$

The rate of work done on the oscillator by the driving force is maximum at resonance ($\omega_1 = \omega_0$). In that case the fluctuation–dissipation theorem reads

$$\frac{c}{2\gamma} + \frac{mb^2}{2\gamma^2} = T \quad \text{(At resonance: } \omega_1 = \omega_0) \qquad (A.12)$$

In steady-state motion, the driving force transfers heat to the medium through "stirring." The theorem holds only when we wait long enough for the steady-state equilibrium to be established. The steady-state temperature $T$ may be different from that in the absence of the driving force.

## A.2.  Coupled Oscillators

### A.2.1.  *Equations of Motion*

Consider a set of linearly coupled driven oscillators, with equations of motion

$$m_\alpha \ddot{q}_\alpha + m_\alpha \omega_\alpha^2 q_\alpha + \gamma_\alpha \dot{q}_\alpha + \sum_\beta \lambda_{\alpha\beta} q_\beta = H_\alpha \quad (\alpha = 1, \ldots, N) \quad (A.13)$$

---

[1]That is, we set the Boltzmann constant to unity ($h = 1$).

where

$$\lambda_{\alpha\beta} = \lambda_{\beta\alpha}, \quad \lambda_{\alpha\alpha} = 0$$
$$H_\alpha = F_\alpha + G_\alpha$$
$$G_\alpha(t) = b_\alpha \cos \nu_\alpha t$$
$$\langle F_\alpha(t) F_\beta(t') \rangle = c_\alpha \delta_{\alpha\beta}(t - t')$$

(A.14)

Now take Fourier transforms:

$$q_\alpha(t) = \frac{1}{2\pi} \int_{-\infty}^{\infty} d\omega e^{-i\omega t} z_\alpha(\omega)$$

$$H_\alpha(t) = \frac{1}{2\pi} \int_{-\infty}^{\infty} d\omega e^{-i\omega t} h_\alpha(\omega)$$

$$h_\alpha = f_\alpha + g_\alpha$$

$$g_\alpha(\omega) = \pi b_\alpha \left[ \delta(\omega + \nu_\alpha) e^{i\varphi_\alpha} + \delta(\omega - \nu_\alpha) e^{-i\varphi_\alpha} \right]$$

$$\langle f_\alpha(\omega') f_\beta(\omega) \rangle = 2\pi c_\alpha \delta_{\alpha\beta} \delta(\omega + \omega')$$

(A.15)

The equations become

$$m_\alpha \left( -\omega^2 + \omega_\alpha^2 - i\omega\kappa_\alpha \right) z_\alpha(\omega) + \sum_\beta \lambda_{\alpha\beta} z_\beta(\omega) = h_\alpha(\omega) \qquad \text{(A.16)}$$

Steady-state solutions can be put in matrix form:

$$z = Qh$$
$$\left( Q^{-1} \right)_{\alpha\beta} = m_\alpha \left( -\omega^2 + \omega_\alpha^2 - i\omega\kappa_\alpha \right) \delta_{\alpha\beta} + \lambda_{\alpha\beta}$$

(A.17)

where

$$\kappa_\alpha = \frac{\gamma_\alpha}{m_\alpha} \qquad \text{(A.18)}$$

We note that

$$Q^*(\omega) = Q(-\omega) \qquad \text{(A.19)}$$

Transient solutions are those of the homogeneous equations (i.e. with $H_\alpha = 0$), but they will decay in time, and will not be considered.

## A.2.2.  Energy Balance

The equations for energy balance can be obtained by multiplying the equations of motion by $\dot{q}_\alpha$ on both sides, and taking ensemble averages:

$$m_\alpha \langle \dot{q}_\alpha \ddot{q}_\alpha \rangle + m_\alpha \omega_\alpha^2 \langle q_\alpha^2 \rangle + \gamma_\alpha \langle \dot{q}_\alpha^2 \rangle + \sum_\beta \lambda_{\alpha\beta} \langle \dot{q}_\alpha q_\beta \rangle = \langle \dot{q}_\alpha H_\alpha \rangle \quad \text{(A.20)}$$

which can be put in the form

$$\frac{d}{dt}(K_\alpha + P_\alpha) = \langle \dot{q}_\alpha H_\alpha \rangle - 2\kappa_\alpha K_\alpha - \sum_\beta \lambda_{\alpha\beta} \langle \dot{q}_\alpha q_\beta \rangle$$

$$K_\alpha(t) = \frac{m_\alpha}{2} \left\langle \dot{q}_\alpha^2(t) \right\rangle \quad \text{(A.21)}$$

$$P_\alpha(t) = \frac{m_\alpha \omega_\alpha^2}{2} \left\langle q_\alpha^2(t) \right\rangle$$

We can make the following identifications:

- Energy input rate from external forces $= \langle \dot{q}_\alpha H_\alpha \rangle \equiv I_\alpha$
- Energy dissipation rate to environment $= 2\kappa_\alpha K_\alpha$
- Energy transfer rate to other oscillators $= \sum_\beta \lambda_{\alpha\beta} \langle \dot{q}_\alpha q_\beta \rangle \equiv R_\alpha$

These quantities contain terms that oscillate with the driving frequencies. We average over an interval longer than the periods of the driving frequencies, and indicate this operation by overhead bar. The average kinetic and potential energies are found to be

$$\bar{K}_\alpha = \frac{1}{4\pi} \sum_\sigma m_\alpha c_\sigma \int_{-\infty}^{\infty} d\omega \, \omega^2 Q_{\alpha\sigma}(\omega) Q_{\beta\sigma}^*(\omega)$$

$$\bar{P}_\alpha = \frac{1}{4\pi} \sum_\sigma m_\alpha \omega_\alpha^2 c_\sigma \int_{-\infty}^{\infty} d\omega \, Q_{\alpha\sigma}(\omega) Q_{\alpha\sigma}^*(\omega) \quad \text{(A.22)}$$

The average energy input rate and average energy transfer rate are given by

$$\bar{I}_\alpha = \frac{c_\alpha}{\pi} \int_0^{\infty} d\omega \, \omega \, \text{Im} \, Q_{\alpha\alpha}(\omega) + \frac{\nu_\alpha b_\alpha^2}{2} \, \text{Im} \, Q_{\alpha\alpha}(\nu_\alpha)$$

$$\bar{R}_\alpha = \frac{1}{\pi} \sum_{\beta,\sigma} \lambda_{\alpha\beta} c_\sigma \int_0^{\infty} d\omega \, \omega \, \text{Im} \left[ Q_{\alpha\sigma}(\omega) Q_{\beta\sigma}^*(\omega) \right] \quad \text{(A.23)}$$

Let $c$ be the diagonal matrix whose diagonal elements are $c_\alpha$. We can write

$$\bar{R}_\alpha = \frac{1}{\pi} \int_0^\infty d\omega\omega \, \text{Im}\left(QcQ^+\lambda\right)_{\alpha\alpha} \tag{A.24}$$

Note that $\text{Tr}(QcQ^+\lambda)$ is real:

$$\left[\text{Tr}\left(QcQ^+\lambda\right)\right]^* = \text{Tr}\left(\lambda QcQ^+\right) = \text{Tr}\left(QcQ^+\lambda\right) \tag{A.25}$$

Thus

$$\sum_\alpha \bar{R}_\alpha = \frac{1}{\pi} \int_0^\infty d\omega\omega \, \text{Im} \, \text{Tr}\left(QcQ^+\lambda\right) = 0 \tag{A.26}$$

This implies that, in steady state, a constant amount of energy is being circulated among the modes.

### A.2.3.   *Fluctuation–dissipation theorem*

The equations of energy balance are

$$\frac{d\bar{K}_\alpha}{dt} + 2\kappa_\alpha \bar{K}_\alpha = \bar{I}_\alpha - \bar{R}_\alpha \tag{A.27}$$

We assume that $d\bar{K}_\alpha/dt = 0$ after a sufficiently long time, and that $\bar{K}_\alpha = T/2$, and obtain the fluctuation–dissipation theorem

$$T\kappa_\alpha = \bar{I}_\alpha - \bar{R}_\alpha \tag{A.28}$$

which is a constraint on the parameters $c_\alpha$ and $\gamma_\alpha$.

### A.2.4.   *Perturbation Theory*

Let $D$ be the diagonal matrix with diagonal elements

$$D_\alpha(\omega) = m_\alpha\left[(\omega^2 - \omega_\alpha^2) + i\omega\kappa_\alpha\right] \tag{A.29}$$

We can expand the matrix $Q$ in terms of the interaction matrix $\lambda$

$$\begin{aligned} Q &= -(D - \lambda)^{-1} = -(1 - D^{-1}\lambda)^{-1}D^{-1} \\ &= -D^{-1} + D^{-1}\lambda D^{-1} - D^{-1}\lambda D^{-1}\lambda D^{-1} + \cdots \end{aligned} \tag{A.30}$$

We cite results of calculations to second order in $\lambda$, as follows.

The input rates are given by

$$\bar{I}_\alpha = \frac{c_\alpha}{2m_\alpha} + I_\alpha^{(2)} + \frac{\kappa_\alpha \nu_\alpha^2 b_\alpha^2}{2m_\alpha} \frac{1}{(\nu_\alpha^2 - \omega_\alpha^2)^2 + \nu_\alpha^2 \kappa_\alpha^2} \qquad (A.31)$$

where

$$I_\alpha^{(2)} = \frac{c_\alpha}{\pi m_\alpha^2} \sum_{\beta \neq \alpha} \frac{\lambda_{\alpha\beta}^2}{m_\beta} \int_0^\infty d\omega$$

$$\times \, \omega^2 \frac{\kappa_\beta \left[ (\omega^2 - \omega_\alpha^2)^2 - (\omega \kappa_\alpha)^2 \right] + 2\kappa_\alpha (\omega^2 - \omega_\alpha^2)(\omega^2 - \omega_\beta^2)}{\left[ (\omega^2 - \omega_\alpha^2)^2 + (\omega \kappa_\alpha)^2 \right]^2 \left[ (\omega^2 - \omega_\beta^2)^2 + (\omega \kappa_\beta)^2 \right]}$$

$$(A.32)$$

The first two terms in $\bar{I}_\alpha$ represent, respectively, direct and indirect effects of the random force. The last term in $\bar{I}_\alpha$ is the average input rate due to the driving force.

The rates of energy transfer are given by

$$\bar{R}_\alpha = \frac{1}{\pi m_\alpha} \sum_{\beta \neq \alpha} \frac{\lambda_{\alpha\beta}^2}{m_\beta} \left( \frac{c_\beta \kappa_\alpha}{m_\beta} - \frac{c_\alpha \kappa_\beta}{m_\alpha} \right)$$

$$\times \int_0^\infty d\omega \omega^2 \frac{1}{\left[ (\omega^2 - \omega_\alpha^2)^2 + \omega^2 \kappa_\alpha^2 \right] \left[ (\omega^2 - \omega_\beta^2)^2 + \omega^2 \kappa_\beta^2 \right]}$$

$$(A.33)$$

Clearly, $\sum_\alpha \bar{R}_\alpha = 0$.

## A.2.5. *Weak-Damping Approximation*

The integrals in $I_\alpha^{(2)}$ and $\bar{R}_\alpha$ are elementary but complicated. To obtain simpler forms, we consider the limit in which all damping coefficients are small:

$$\gamma_\alpha \to 0 \qquad (A.34)$$

Using the formulas

$$\frac{\epsilon}{x^2 + \epsilon^2} \xrightarrow[\epsilon \to 0]{} \pi\delta(x)$$

$$\frac{\epsilon^3}{(x^2 + \epsilon^2)^2} \xrightarrow[\epsilon \to 0]{} \frac{\pi}{2}\delta(x) \tag{A.35}$$

we obtain

$$\bar{I}_\alpha \approx \frac{c_\alpha}{2m_\alpha} \left\{ 1 + \sum_{\beta \neq \alpha} \frac{\lambda_{\alpha\beta}^2}{m_\alpha m_\beta} \left( 1 - \frac{\kappa_\beta}{2\kappa_\alpha} \right) \frac{1}{\left( \omega_\alpha^2 - \omega_\beta^2 \right)^2} \right\}$$

$$+ \frac{\kappa_\alpha \nu_\alpha^2 b_\alpha^2}{2m_\alpha} \frac{1}{\left( \nu_\alpha^2 - \omega_\alpha^2 \right)^2 + \nu_\alpha^2 \kappa_\alpha^2}$$

$$\bar{R}_\alpha \approx \sum_{\beta \neq \alpha} \frac{\lambda_{\alpha\beta}^2}{2m_\alpha m_\beta} \left( \frac{c_\beta \kappa_\alpha}{m_\beta} - \frac{c_\alpha \kappa_\beta}{m_\alpha} \right) \left( \frac{1}{\kappa_\alpha} + \frac{1}{\kappa_\beta} \right) \frac{1}{\left( \omega_\alpha^2 - \omega_\beta^2 \right)^2}$$

$$\tag{A.36}$$

We have assumed that there are no degeneracies among the frequencies $\omega_\alpha$.

## A.3.  Model of Protein Dynamics

We consider a protein molecule near the molten globule state, where there is a clear distinction between the surface of the molecule and its interior. The surface is in contact with water, and subject to random impacts from water molecules. The interior feels the random forces only indirectly, through their couplings to the surface modes. We assume that the molecules is set into resonant vibration with water networks surrounding the molecule, and there is resonant transfer of organized energy. The frequencies involved should be of the order of 10–100 GHz.

In the high-dimensional configuration space of the protein molecule, the potential energy has many local minima. These arise from the fact that, as the spatial structure of the molecule varies, different chemical bonds "snap" into (or unsnap out of) positions. During the folding process, the molecule follows a path in this configuration space, and the potential energy goes through a series

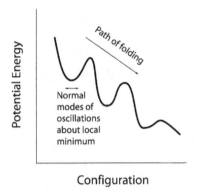

Fig. A.1 Potential energy of protein molecule along the folding path. The molecule is momentarily trapped in a local minimum. Its structure can be described through normal modes of oscillations about that minimum. As the molecule moves on to the next minimum, the normal modes change. The folding process can be decribed as a sequence of normal modes.

of local minima. We are interested in motions much faster than the folding process, so that the molecule appears to oscillate about the local minima. This is symbolically depicted in Fig. A.1.

We assume that normal frequencies with respect to that minimum are non-degenerate:

$$\omega_1 < \omega_2 < \cdots < \omega_N \tag{A.37}$$

with associated masses $m_1, \ldots, m_N$. As the folding process evolves, these parameters change on a slower time scale.

- The mode $\omega_1$ corresponds to motion on the largest length scale, associated with overall deformations of the molecule. We assume that this is the only mode driven by external force, and we assume it is driven at resonance $\nu_1 = \omega_1$. In addition, this mode is subjected to random forces, and dissipates energy.
- The mode $\omega_N$ corresponds to motions on the smallest length scale — the surface modes. It is not being driven, but is subjected to random forces, and dissipated energy.
- All other modes are internal modes. They are not directly driven, not directly subjected to random forces, and do not directly dissipate energy. However, Brownian motions are induced through couplings to $\omega_1$ and $\omega_N$.

Thus,

$$
\begin{aligned}
b_\alpha &= 0 \quad (\alpha = 2, \ldots, N) \\
c_\alpha &= \gamma_\alpha = 0 \quad (\alpha = 2, \ldots, N-1)
\end{aligned}
\tag{A.38}
$$

Apart from the frequencies and the masses, the model parameters are

- Driving force: $b_1$
- Random forces: $c_1$, $c_N$
- Damping constants: $\gamma_1, \gamma_N$

We use perturbation theory in the weak-damping approximation. From (A.36), we see that

$$
\bar{I}_\alpha = \bar{R}_\alpha = 0 \quad (\alpha = 2, \ldots, N-1)
\tag{A.39}
$$

We need only consider the modes $\alpha = 1, N$:

$$
\begin{aligned}
\bar{I}_1 &= \frac{c_1}{2m_1} + \frac{b_1^2}{2\gamma_1} + \frac{c_1}{m_1}\left[\left(1 - \frac{\kappa_N}{2\kappa_1}\right)B + F\right] \\
\bar{I}_N &= \frac{c_N}{2m_N} + \frac{c_N}{m_N}\left[\left(1 - \frac{\kappa_1}{2\kappa_N}\right)B + F'\right] \\
\bar{R}_1 &= \left(\frac{c_N\gamma_1}{m_N^2} - \frac{c_1\gamma_N}{m_1^2}\right)\left(\frac{m_1}{\gamma_1} + \frac{m_N}{\gamma_N}\right)B \\
\bar{R}_N &= -\bar{R}_1
\end{aligned}
\tag{A.40}
$$

where $B$, $F$, $F'$ are dimensionless quantities given by

$$
\begin{aligned}
B &= \frac{\lambda_{1N}^2}{2m_1 m_N\left(\omega_1^2 - \omega_N^2\right)^2} \\
F &= \sum_{\beta=2}^{N-1} \frac{\lambda_{1\beta}^2}{2m_1 m_\beta} \frac{1}{\left(\omega_1^2 - \omega_\beta^2\right)^2} \\
F' &= \sum_{\beta=2}^{N-1} \frac{\lambda_{N\beta}^2}{2m_N m_\beta} \frac{1}{\left(\omega_N^2 - \omega_\beta^2\right)^2}
\end{aligned}
\tag{A.41}
$$

## A.4.   Fluctuation–Dissipation Theorem

The fluctuation–dissipation theorem (A.28) is automatically satisfied for $\alpha = 2, \ldots, N - 1$. For $\alpha = 1, N$, we have

$$
\begin{aligned}
T\kappa_1 &= \bar{I}_1 - \bar{R}_1 \\
T\kappa_N &= \bar{I}_N + \bar{R}_1
\end{aligned}
\tag{A.42}
$$

These are to be satisfied order-by-order in perturbation theory. The unperturbed conditions are

$$
\begin{aligned}
2T\kappa_1 &= \frac{c_1}{m_1} + \frac{b_1^2}{m_1 \kappa_1} \\
2T\kappa_N &= \frac{c_N}{m_N}
\end{aligned}
\tag{A.43}
$$

The second-order conditions are

$$
\begin{aligned}
0 &= \frac{c_1}{m_1} \left[ \left( 1 - \frac{\kappa_N}{2\kappa_1} \right) B + F \right] - \left( \frac{c_N \kappa_1}{m_N} - \frac{c_1 \kappa_N}{m_1} \right) \left( \frac{1}{\kappa_1} + \frac{1}{\kappa_N} \right) B \\
0 &= \frac{c_N}{m_N} \left[ \left( 1 - \frac{\kappa_N}{2\kappa_1} \right) B + F' \right] + \left( \frac{c_N \kappa_1}{m_N} - \frac{c_1 \kappa_N}{m_1} \right) \left( \frac{1}{\kappa_1} + \frac{1}{\kappa_N} \right) B
\end{aligned}
\tag{A.44}
$$

From these four conditions, we can determine $c_1, c_N, \gamma_1, \gamma_N$ in terms of the other parameters.

We simplify the notation by introducing

$$
\begin{aligned}
m &= m_1, \quad c = c_1, \quad \gamma = \gamma_1, \quad b = b_1 \\
r &= \frac{m_N}{m_1}, \quad s = \frac{\kappa_N}{\kappa_1}, \quad t = \frac{c_N}{c_1}
\end{aligned}
\tag{A.45}
$$

The conditions to be satisfied are then

$$
\begin{aligned}
2T &= \frac{c}{\gamma} + \frac{mb^2}{\gamma^2} \\
2T &= \frac{c}{\gamma} \frac{t}{rs} \\
0 &= 1 - \frac{s}{2} - \left( \frac{t}{r} - s \right) \left( 1 + \frac{1}{s} \right) + \frac{F}{B} \\
0 &= 1 - \frac{s}{2} + \left( 1 - \frac{sr}{t} \right) \left( 1 + \frac{1}{s} \right) + \frac{F'}{B}
\end{aligned}
\tag{A.46}
$$

These are to be solved for $s, t, c, \gamma$.

Now assume that the global vibration has a much higher mass than that of the surface mode:

$$m_1 \gg m_N \tag{A.47}$$

Thus

$$r \ll 1, \quad \frac{F'}{B} \gg 1, \quad \frac{F}{B} \gg 1 \tag{A.48}$$

The last two inequalities hold because $F, F'$ are sums of the order of $N$ positive terms, each comparable to $B$.

With these approximations, the conditions reduce to

$$
\begin{aligned}
2T &= \frac{c}{\gamma} + \frac{mb^2}{\gamma^2} \\
2T &= \frac{c}{\gamma} \frac{t}{rs} \\
s &= \frac{2t}{r} \\
s &= \frac{2F'}{B}
\end{aligned} \tag{A.49}
$$

Solving these give

$$
\begin{aligned}
s &= \frac{m_1}{m_N} \frac{\gamma_N}{\gamma_1} = \frac{2F'}{B} \\
t &= \frac{c_N}{c_1} = \frac{rF'}{B}\left(1 + \frac{F}{F'}\right) \\
\gamma &= \gamma_1 = \sqrt{\frac{mb^2}{2T}} \sqrt{\frac{F + F'}{F - F'}} \\
c &= c_1 = \frac{2\sqrt{2mb^2 T}}{\sqrt{(F/F')^2 - 1}}
\end{aligned} \tag{A.50}
$$

## A.5.   The Cascade Time

The energy input rate is given by

$$\bar{I}_1 = \frac{c}{2m} + \frac{b^2}{2\gamma} + \frac{c}{m}\left(F - \frac{s}{2}B\right) \approx 2\sqrt{\frac{2b^2 T}{m}} F' \sqrt{\frac{F - F'}{F + F'}} \tag{A.51}$$

In a steady state, this must equal the amount of energy dissipated per unit time, and the energy being dissipated should be of the order of the thermal energy $T$. Thus, the inverse cascade time is of the order of $\tau^{-1} = \bar{I}_1/T$, hence

$$\tau = \frac{\tau_0}{F'} \sqrt{\frac{F + F'}{F - F'}} \tag{A.52}$$

where

$$\tau_0 = \frac{1}{2} \sqrt{\frac{mT}{2b^2}} \tag{A.53}$$

is the characteristic time of the energy cascade. We minimize $\tau$ by varying the system parameters, i.e. by changing the structure of the molecule.

## A.6.   Numerical Example

For illustration, we work out an example in which the structure is specified by only one adjustable parameter

$$\xi = \frac{N\omega_1}{\omega_N} \tag{A.54}$$

We shall see that the cascade time has a minimum with respect to $\xi$, and that minimum would correspond to the equilibrium structure of the molecule.

First, let us rewrite

$$B = \frac{ru_1}{(1 - f_1)^2}$$

$$F = r \sum_{\beta=2}^{N-1} \frac{m_N}{m_\beta} \frac{v_\beta}{(f_1 - f_\beta)^2} \tag{A.55}$$

$$F' = \sum_{\beta=2}^{N-1} \frac{m_N}{m_\beta} \frac{u_\beta}{(1 - f_\beta)^2}$$

where

$$f_\beta = \left( \frac{\omega_\beta}{\omega_N} \right)^2$$

$$u_\beta = \frac{1}{2} \left( \frac{\lambda_{N\beta}}{m_N \omega_N^2} \right)^2 \tag{A.56}$$

$$v_\beta = \frac{1}{2} \left( \frac{\lambda_{1\beta}}{m_N \omega_N^2} \right)^2$$

Now choose frequencies such that they are uniformly spaced between $\omega_1$ and $\omega_N$, with

$$\frac{\omega_\alpha}{\omega_N} = \frac{\xi}{N} + (\alpha - 1)\Delta$$

$$\Delta = \frac{1}{N-1} \left( 1 - \frac{\xi}{N} \right) \tag{A.57}$$

Thus

$$f_n = \left[ \frac{\xi}{N} + (n-1)\Delta \right]^2 \tag{A.58}$$

The interactions are taken to be

$$v_\beta = 1$$

$$u_\beta = \frac{1}{N} \tag{A.59}$$

The interactions of the high-frequency mode $u_\beta$ are assumed to be smaller, because the mode should be less affected by the folding process.

We take

$$r = \frac{m_N}{m_1} = \frac{1}{N} \tag{A.60}$$

and assume that masses decrease linearly with $\alpha$:

$$m_\alpha = m - \frac{\alpha - 1}{N-1}(m - m_N) = m \left[ 1 - \frac{\alpha - 1}{N} \right] \tag{A.61}$$

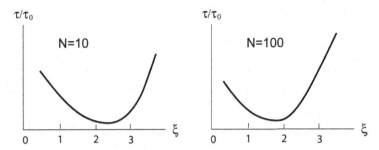

Fig. A.2  Energy cascade time as a function of a structural parameter. According to our proposed principle, the minimum determines the equilibrium structure.

Then

$$F = \frac{1}{N} \sum_{\beta=2}^{N-1} \frac{1}{(N+1-\beta)(f_1 - f_\beta)^2}$$

$$F' = \frac{1}{N} \sum_{\beta=2}^{N-1} \frac{1}{(N+1-\beta)(1 - f_\beta)^2}$$

$$(A.62)$$

We calculate $\tau/\tau_0$, and vary $\xi$ to search for a minimum. The results are shown in Fig. A.2, and summarized in the table below:

| $N$ | $\xi$ | $\omega_1/\omega_N$ |
|-----|-------|---------------------|
| 10  | 2.3   | 0.23                |
| 100 | 1.9   | 0.019               |

The values of $\tau$ cannot be compared for the two cases, since the $N$ dependence of $\tau_0$ (in particular $b$) is not specified in this model.

# Index